AI and the Boardroom

Insights into Governance, Strategy, and the Responsible Adoption of AI

Rohan Sharma

Apress®

AI and the Boardroom: Insights into Governance, Strategy, and the Responsible Adoption of AI

Rohan Sharma
Murrieta, CA, USA

ISBN-13 (pbk): 979-8-8688-0795-4   ISBN-13 (electronic): 979-8-8688-0796-1
https://doi.org/10.1007/979-8-8688-0796-1

Copyright © 2024 by Rohan Sharma

This work is subject to copyright. All rights are reserved by the Publisher, whether the whole or part of the material is concerned, specifically the rights of translation, reprinting, reuse of illustrations, recitation, broadcasting, reproduction on microfilms or in any other physical way, and transmission or information storage and retrieval, electronic adaptation, computer software, or by similar or dissimilar methodology now known or hereafter developed.

Trademarked names, logos, and images may appear in this book. Rather than use a trademark symbol with every occurrence of a trademarked name, logo, or image we use the names, logos, and images only in an editorial fashion and to the benefit of the trademark owner, with no intention of infringement of the trademark.

The use in this publication of trade names, trademarks, service marks, and similar terms, even if they are not identified as such, is not to be taken as an expression of opinion as to whether or not they are subject to proprietary rights.

While the advice and information in this book are believed to be true and accurate at the date of publication, neither the authors nor the editors nor the publisher can accept any legal responsibility for any errors or omissions that may be made. The publisher makes no warranty, express or implied, with respect to the material contained herein.

    Managing Director, Apress Media LLC: Welmoed Spahr
    Acquisitions Editor: Shivangi Ramachandran
    Development Editor: James Markham
    Editorial Assistant: Jessica Vakili

Cover designed by eStudioCalamar

Distributed to the book trade worldwide by Springer Science+Business Media New York, 1 New York Plaza, Suite 4600, New York, NY 10004-1562, USA. Phone 1-800-SPRINGER, fax (201) 348-4505, e-mail orders-ny@springer-sbm.com, or visit www.springeronline.com. Apress Media, LLC is a California LLC and the sole member (owner) is Springer Science + Business Media Finance Inc (SSBM Finance Inc). SSBM Finance Inc is a **Delaware** corporation.

For information on translations, please e-mail booktranslations@springernature.com; for reprint, paperback, or audio rights, please e-mail bookpermissions@springernature.com.

Apress titles may be purchased in bulk for academic, corporate, or promotional use. eBook versions and licenses are also available for most titles. For more information, reference our Print and eBook Bulk Sales web page at http://www.apress.com/bulk-sales.

Any source code or other supplementary material referenced by the author in this book is available to readers on GitHub. For more detailed information, please visit https://www.apress.com/gp/services/source-code.

If disposing of this product, please recycle the paper

Table of Contents

About the Author xxxi

Acknowledgments xxxiii

Chapter 1: Introduction...1
   Summary..4

Chapter 2: AI Governance 5
   Introduction...5
   Drivers for AI Governance 6
      Driver 1: Legal and Regulatory Compliance 6
      Driver 2: Ensuring Accountability..................................6
   AI Governance: Approaches for Federal and State Governments..........7
   AI Governance: Approach for Organizations............................8
   An Integrated Framework for AI Deployment and Governance 9
   The AI Deployment and Governance Framework 10
      1. Discover...11
      2. Create...11
      3. Execute..12
      4. Operate..13
   AI Lifecycle Governance 13
      Governance Framework 13

TABLE OF CONTENTS

Strategic Implementation of AI Deployment and Governance 14
    1. Establish Clear Governance Policies 14
    2. Invest in Robust Data Management 15
    3. Foster Cross-Functional Collaboration 15
    4. Implement Continuous Monitoring and Improvement 15
    5. Prioritize Ethical Considerations 15
    6. Align AI Initiatives with Business Strategy 16
Crafting a Robust Operational AI Governance Framework 16
    Design Phase: Laying the Foundation 17
    Build Phase: Constructing the Solution 18
    Run Phase: Operationalizing AI 19
AI Governance Checklist for Board and C-Suite Executives 20
    Scoring System 20
    Environmental Layer: Setting the Stage for AI Governance 20
    Organizational Layer: Aligning AI with Business Strategy 21
    AI System Layer: Ensuring Operational Excellence 22
    Accountability and Compliance 24
    Continuous Improvement and Knowledge Flow 25
Total Scoring 25
Interpretation 25
Summary 26

Chapter 3: AI Regulation 27
Approaches to AI Regulation 28
Government Regulations 29
European Union 30
The United States 30
Role of Government in Regulating Technology 31

TABLE OF CONTENTS

A Self-Regulatory Framework 31
    Checklist for AI Regulations Considerations for Board and C-Suite 32
Summary 36

Chapter 4: AI Privacy 37
Introduction 37
Checklist for AI Privacy Considerations for Board and C-Suite 40
    Scoring System 40
    General Compliance 40
    Consent and Data Subject Rights 41
    Data Security and Breaches 42
    Risk Assessment and Governance 42
    Privacy by Design and Data Transfers 43
    Organizational Culture and Training 44
    Ongoing Evaluation and Improvement 44
Total Scoring 44
Interpretation 45
Summary 45

Chapter 5: AI Copyright and Intellectual Property 47
IP and Copyright Laws and How They Apply to AI-Generated Content 48
The Million-Dollar Question Is, Does Your AI-Generated Work Get
Copyright Protection Right Now? 48
Ongoing AI Innovations and Their IP Dilemmas 49
    AlphaGo by DeepMind 50
    GPT-3 by OpenAI 50
    Artbreeder's AI Creations 50
    Chatbots and Fair Use 50
    IBM Watson's Patent Insight Tool 50

TABLE OF CONTENTS

Defining the AI IP and Copyright Landscape 51
Proposed US and EU Copyright Laws and Regulations 53
AI Copyright and IP Consideration Checklist for Board and C-Suite 54
    General Compliance 54
    Authorship and Inventorship 54
    Data Usage and Copyright 55
    Predictive Capabilities and IP Strategy 55
    Legal Framework and Compliance 55
    Data Security and Breach Response 56
    Continuous Monitoring and Improvement 56
Scoring and Interpretation 56
Threshold for Passing 57
Summary 57

Chapter 6: AI Strategy 59
Defining Your AI Strategy 59
Avoid Overemphasis on Cost Savings 60
Communicating a Clear AI Strategy Elevates Market Value 61
Mindset for a Robust AI Strategy 61
    1. Setting the Legal and Compliance Groundwork 63
    2. Strategic Framework for AI Implementation 64
    3. Prioritizing Key AI Initiatives 64
    4. Exploration and Innovation in AI 65
    5. AI Strategy Roadmapping 65
    6. Ethical AI Deployment 66
Navigating the Buy vs. Build Decision in AI Strategy 66
    Tier 1: Basic LLM Integration 67
    Tier 2: Customized LLM Implementation 68

TABLE OF CONTENTS

Tier 3: Advanced LLM Pipelines 68
Tier 4: Enterprise-wide LLM Adoption 69
AI Strategy Consideration Checklist for Board and C-Suite 69
   Scoring System 69
   Setting the Legal and Compliance Groundwork 70
   Strategic Framework for AI Implementation 70
   Prioritizing Key AI Initiatives 71
   Exploration and Innovation in AI 72
   AI Strategy Roadmapping 72
   Ethical AI Deployment 73
   Leadership and Communication 73
   Commitment and Long-Term Investment 74
   Scoring and Interpretation 75
   Threshold for Passing 75
Summary 75

Chapter 7: AI Operating Model 77

The AI Operating Model Framework 78
   1. Data Providers 79
   2. IT Operations 80
   3. Data Science 80
   4. Data and AI Governance 81
   5. Data Consumers 81
   6. AI Service Delivery Management 81
Implementing the AI Operating Model 82
   1. Establish Clear Roles and Responsibilities 82
   2. Develop Robust Data Governance Policies 83
   3. Foster Collaboration Across Teams 83
   4. Invest in Technology and Infrastructure 83

TABLE OF CONTENTS

    5. Monitor and Optimize Performance 83

    6. Ensure Ethical AI Practices 84

Strategic Implications for Executives and Boards 84

    1. Alignment with Strategic Goals 84

    2. Enhanced Decision-Making 85

    3. Operational Efficiency 85

    4. Risk Mitigation 85

    5. Competitive Advantage 85

Who Owns Data and AI Budgets? 85

Common Challenges in AI Operating Model Delivery at Scale 86

AI Operating Model Checklist for Leaders 88

    Scoring System 88

    Alignment with Business Use Cases 88

    Agile and Iterative Approach 89

    Collaboration and Integration 89

    Accountability and Governance 89

    Data and IT Infrastructure 90

    Data Science and AI Development 90

    Governance and Ethical Considerations 90

    Service Delivery and Performance Monitoring 91

    Ethical AI Practices 91

    Strategic Implementation 92

    Total Scoring 93

    Interpretation 93

    Threshold for Passing 93

Summary 94

Chapter 8: Determining AI Maturity for Your Organization 95

Levels of AI Maturity 96
    Level 0: Foundational 96
    Level 1: Basic Retrieval Augmentation 96
    Level 2: Intermediate Retrieval Augmentation 96
    Level 3: Advanced Retrieval Augmentation 96
    Level 4: Advanced Retrieval Augmentation with FFT 97
    Level 5: Orchestrated Agentic Systems 97
    Level 6: Multiagent Systems and Workflow Orchestration 97

Strategic Implications for Executives and Boards 98
    1. Invest in Data Infrastructure 98
    2. Develop Incremental AI Capabilities 98
    3. Foster Cross-Functional Collaboration 98
    4. Prioritize Responsible AI Practices 99
    5. Continuously Evaluate and Optimize 99

Summary 104

Chapter 9: Structuring AI Teams for Success: Models for Scaling AI Operations 105

Models for Structuring AI Teams 105
    1. Functional Model 106
    2. Centralized Model 106
    3. Decentralized Model 107
    4. Factory Model 108
    5. Center of Excellence (CoE) 108
    6. Consulting Model 109

Choosing the Right Model 109
Strategic Implementation of AI Structures 111

TABLE OF CONTENTS

AI Team Structure Success Checklist for Board and C-Suite 113
- Scoring System 113
- Functional Model 113
- Centralized Model 113
- Decentralized Model 114
- Factory Model 114
- Center of Excellence (CoE) 114
- Consulting Model 115
- Strategic Considerations 115
- Governance and Coordination 115
- Collaboration and Innovation 116
- Strategic Implementation Steps 116
- Monitoring and Optimization 116
- Total Scoring 117
- Scoring and Interpretation 117
- Threshold for Passing 117

Summary 117

Chapter 10: AI Partnerships and Alliances 119
AI Partnerships and Alliances Checklist for Board and C-Suite 122
- Scoring System 122

Checklist Questions 123
- Strategic Alignment 123
- Technical Proficiency 123
- Ecosystem and Collaboration 124
- Change Management and Talent 124
- Ethical Considerations 125
- Risk Management and Governance 125

Total Scoring 125

Scoring and Interpretation 126
Threshold for Passing 126
Summary.. 126

Chapter 11: AI Budgets and Investments 127
The Trends in AI Investment in the Organizations 127
    AI Budgets and Investments: Checklist for Board and C-Suite................... 129
    Checklist for AI Budgets and Investments 129
    Total Scoring 132
    Scoring and Interpretation.. 132
    Threshold for Passing 133
Summary.. 133

Chapter 12: AI Change Management 135
Tip 1: Set Ambitious Goals, Start with Small Steps.............................. 135
Tip 2: Prioritize Human-Centered Design 136
Case Study: Enhancing Emergency Response Times........................... 136
Cost Efficiency and Growth: The Dual Benefits of AI 136
OCM: Emphasizing the Human Aspect of AI Change Management 137
Leadership's Role in AI-Driven Change 137
Addressing Resistance to Change 137
Effective Communication and Training... 138
PPM: The Structural Backbone of AI Change Management................. 138
    PPM's Contribution to Successful AI Adoption... 138
Strategic Considerations for AI Adoption: Ensuring Effective
Implementation 139
    Key Considerations for AI Adoption.. 139
    1. Data Ownership and Licensing... 139
    2. Input Validation and Sanitization 140

    3. Model Robustness 140
    4. Data Privacy and Compliance 140
    5. Technical Challenges 140
    6. Ethical Considerations 141
    7. Technical Expertise 141
    8. Problem–Solution Fit 141
    9. Data Availability and Quality 141
    10. Common AI Applications 142
    Strategic Directions for AI Adoption 142
AI Change Management Checklist for Board and C-Suite 145
    Scoring System 145
    General Compliance 145
    Change Management Framework 145
    Stakeholder Engagement 146
    Governance and Leadership 146
    Resource Allocation 146
    Training and Development 147
    Monitoring and Evaluation 147
    Risk Management 147
    Cultural Adaptation 148
    Financial Management 148
    Technology Integration 148
    Measuring Success 149
    Total Scoring 149
    Scoring and Interpretation 149
    Threshold for Passing 150
  Summary 150

Chapter 13: AI KPIs and OKRs: Measuring Success and Maximizing Impact 151

The Importance of Metrics in AI Initiatives 151

Setting the Context with KPIs and Metrics 152

Types of AI Metrics 153

Selecting the Right Metrics 154

Essential KPIs for AI Projects 156

    Operational Efficiency 157

    Customer Satisfaction 157

    Revenue Growth 158

Quantifying AI's ROI 159

    Cost Savings vs. Investment Costs 159

Challenges in Measuring AI Success 160

Strategies for Overcoming Measurement Challenges 163

Demonstrating AI's Business Value 163

Guiding Future AI Strategies 164

Future Planning 165

AI OKR and KPI Checklist for Board and C-Suite 165

    Scoring System 165

    General Considerations 165

    Efficiency Metrics 166

    Accuracy Metrics 166

    Performance Metrics 167

    Financial Impact Metrics 167

    Customer Satisfaction Metrics 168

    Strategic Implementation 168

    Continuous Improvement 169

TABLE OF CONTENTS

    Total Scoring 169
    Scoring and Interpretation 169
    Threshold for Passing 170
  Summary 170

Chapter 14: AI Partnerships and Strategic Alliances 171
  The Strategic Importance of AI Partnerships 171
  Key Elements of Effective AI Partnerships 172
    1. Deep Collaboration 172
    2. Scalability, Interoperability, and Reusability 173
    3. Maintaining Control and Flexibility 174
  Getting Started with AI Partnerships 175
  Summary 176

Chapter 15: AI Talent Strategy 177
  The AI Talent Landscape 177
  Attracting AI Talent 178
    1. Define a Clear Value Proposition 178
    2. Leverage Untapped Talent Pools 179
    3. Customize Recruiting Processes 179
    4. Anchor Hires 179
  Developing AI Talent 179
    1. Reskilling Programs 180
    2. Continuous Learning Culture 180
    3. Structured Career Paths 180
    4. Communities of Practice 180
  Retaining AI Talent 181
    1. Purpose-Driven Work 181
    2. Integration into the Organization 181

TABLE OF CONTENTS

- 3. Flexible Work Arrangements 181
- 4. Recognition and Rewards 181

Strategic Considerations for AI Talent Management 182
- 1. Data-Driven Decisions 182
- 2. Collaborative Ecosystem 182
- 3. Ethical and Responsible AI 182
- 4. Long-Term Vision 183

Case Study: Booz Allen Hamilton 183
- 1. Early Adoption and Centralized Teams 183
- 2. Proactive Talent Mapping 183
- 3. Partnerships with Educational Institutions 183
- 4. Comprehensive Training Programs 184

Moving Forward: Implementing an AI Talent Strategy 184
- 1. Establish a Steering Committee 184
- 2. Develop a Strategic Talent Playbook 185
- 3. Conduct a Talent Audit 185
- 4. Assign Dedicated Relationship Managers 185
- 5. Foster a Culture of Innovation 185
- 6. Emphasize Ethical AI 185
- 7. Monitor and Adapt 186

Generative AI: Redefining Leadership 186
- Chief Executive Officer (CEO) 187
- Chief Operating Officer (COO) 188
- Chief Information, Technology, and Data Officers (CIOs, CTOs, CDOs) 191
- Chief Legal and Privacy Officers 192
- Chief Product Officer 192
- Chief Marketing Officer 193

XV

TABLE OF CONTENTS

    Chief Financial Officer 193

    Chief Human Resources Officer 194

  AI Talent Strategy Checklist for Board and C-Suite 194

    Scoring System 194

    Attracting AI Talent 195

    Developing AI Talent 196

    Retaining AI Talent 196

    Strategic Considerations for AI Talent Management 197

    Implementation and Monitoring 198

    Total Scoring 200

    Scoring and Interpretation 200

    Threshold for Passing 200

  Summary 201

Chapter 16: AI Monetization: Strategies for Profitable Innovation 203

  AI-as-a-Service (AIaaS) 204

    1. Subscription Models 204

    2. Pay-Per-Use Models 204

    3. Freemium to Premium 204

  Custom AI Solutions 204

    1. Enterprise Customization 205

    2. Confidentiality and Data Security 205

    3. Third-Party Service Providers 205

  AI-Powered Products and Services 205

    1. Enhancing Existing Products 205

    2. Developing New Offerings 206

    3. Market Differentiation 206

TABLE OF CONTENTS

Legal and Ethical Considerations 206
   1. Data Privacy and Security...206
   2. Intellectual Property (IP) 206
   3. Liability and Indemnification 207
Insurance as a Risk Management Tool...207
   1. Bridging Liability Gaps...207
   2. Customized Policies 207
Future Trends in AI Monetization..207
   1. Open Source AI 208
   2. Data As a Competitive Advantage...208
   3. AI Ecosystems 208
Case Study: OpenAI..208
   1. Subscription and Pay-Per-Use Models 208
   2. Token-Based Pricing 209
   3. Freemium Model 209
AI Monetization Consideration Checklist for Board and C-Suite 209
   Scoring System 209
   Common Themes..210
   Monetization Models 211
   Total Scoring...212
Scoring and Interpretation 213
   Threshold for Passing..213
Summary..213

Chapter 17: Aligning AI Investments with Business Problems 215
Understanding the Problem–Investment Matrix 215
   Tier 1: Fundamental AI Applications 216
   Tier 2: Departmental Enhancements 216

xvii

TABLE OF CONTENTS

    Tier 3: Advanced Analytical Tools..217
    Tier 4: Enterprise-wide Transformation 217
  Summary..223

Chapter 18: AI Cybersecurity...225
  Understanding Organizational AI Security Posture 225
  Core LLM Security Concerns..227
  Strategies to Address AI Security Concerns..228
  Building a Resilient AI Cybersecurity Framework 229
  AI Cybersecurity Checklist for Board and C-Suite..230
    Scoring System 230
    Hardened Infrastructure 230
    Access Control..231
    Data Protection 231
    Customer Trust and Transparency 232
    Responsible AI 232
    Identity and Input Validation 233
    Data Handling Risks 233
    Security of AI Models..234
    Continuous Monitoring and Testing 234
    Training and Awareness 234
    Incident Response Planning 235
    Collaboration with Security Experts 235
    Total Scoring...236
    Scoring and Interpretation...236
    Threshold for Passing...236
  Summary..236

TABLE OF CONTENTS

Chapter 19: Scaling AI Operations: Designing Effective Enterprise Infrastructure 237

Technology Infrastructure for AI 238

Investments Required to Support AI Initiatives 239

Human Resources and Expertise 239

Integration Systems for Expanded AI Scope 240

Harnessing Unique, Task-Specific Enterprise Data 240

    Checklist for "Scaling AI Operations: Designing Effective Enterprise Infrastructure" 241

    1. Technology Infrastructure for AI 241

    2. Investments Required to Support AI Initiatives 242

    3. Human Resources and Expertise 243

    4. Integration Systems for Expanded AI Scope 243

    5. Harnessing Unique, Task-Specific Enterprise Data 244

    Total Scoring 244

    Scoring and Interpretation 245

    Threshold for Passing 245

Chapter 20: Building Robust AI Infrastructure for Enterprise Success 247

What Are Large Language Models? 247

Model Architecture and Training 248

LLM Security and Risks 249

LLM Applications 249

Optimizing LLM Performance 249

LLM Evaluation Metrics 250

Challenges and Future Directions 251

    Checklist for "Building Robust AI Infrastructure for AI Success" 252

    Technology Infrastructure for AI 252

TABLE OF CONTENTS

Investments Required to Support AI Initiatives 253
Human Resources and Expertise 254
Integration Systems for Expanded AI Scope 255
Harnessing Unique, Task-Specific Enterprise Data 255
Technical Understanding of Generative AI for Boards and Executives 256
Total Scoring and Interpretation 257
Scoring Threshold 257
Minimum Score to Pass 257

Summary 258

Chapter 21: Architecting AI Solutions: A Blueprint for Generative AI 259

The Generative AI Reference Architecture 259
   1. User Experience (U/X) 260
   2. Prompt Engineering 261
   3. Retrieval Augmentation Generation (RAG) 261
   4. Serving and Orchestrating Models 261
   5. Adaptation and Tuning 262
   6. Evaluation and Observability 262
   7. MLOps Orchestration 262
   8. Security, Privacy, and Compliance 263
   9. Governance and Responsible AI 263
   10. Enterprise Integration 264

Strategic Implications for Businesses 264
   1. Enhanced Innovation 264
   2. Operational Efficiency 265
   3. Competitive Advantage 265
   4. Risk Mitigation 265
   5. Scalability 265

Building Trustworthy AI Systems 265
    Characteristics of Trustworthy AI Systems 266
Checklist for Architecting AI Solutions: A Blueprint for Generative AI 268
    Scoring System 269
    User Experience (UX) 269
    Prompt Engineering 269
    Retrieval Augmentation Generation (RAG) 270
    Serving and Orchestrating Models 270
    Adaptation and Tuning 270
    Evaluation and Observability 271
    MLOps Orchestration 271
    Security, Privacy, and Compliance 271
    Governance and Responsible AI 272
    Enterprise Integration 272
    Total Scoring 272
    Scoring and Interpretation 272
    Threshold for Passing 273
Summary 273

Chapter 22: AI Risk Categorization 275

Operational Risks 275
    1. Unpredictable AI Behavior 276
    2. Data Quality and Integrity 276
    3. Dependency on External Vendors 276
Legal and Regulatory Risks 276
    1. Liability and Accountability 277
    2. Data Privacy and Protection 277
    3. Compliance with Industry Standards 277

TABLE OF CONTENTS

Ethical and Social Risks 277
- 1. Bias and Discrimination 277
- 2. Transparency and Explainability 278
- 3. Impact on Employment 278

Checklist for AI Risk Consideration for Board and C-Suite 278
- Privacy 278
- Security 279
- Fairness 279
- Transparency and Explainability 280
- Safety and Performance 280
- Third-Party Risks 281
- Legal and Regulatory Compliance 281
- Organizational and Cultural Integration 282

Strategies for Mitigating AI Risks 282
- 1. Robust Risk Management Framework 283
- 2. Continuous Monitoring and Auditing 283
- 3. Clear Legal Agreements 283
- 4. Data Governance and Quality Control 284
- 5. Transparent and Explainable AI 284
- 6. Ethical AI Guidelines 284
- 7. Workforce Reskilling and Transition 284

Case Study: Managing AI Risks in Financial Services 284
- 1. Risk Assessment and Mitigation 285
- 2. Legal and Compliance Strategy 285
- 3. Continuous Monitoring and Adjustment 285
- 4. Transparency and Customer Trust 285

Summary 286

TABLE OF CONTENTS

Chapter 23: Strategic AI Risk Management & Quantification 287
  1. Map 288
  2. Measure 288
  3. Manage 288
  4. Govern 288
  Key Risk Categories and Mitigation Strategies 289
   1. Output Quality 289
   2. Data Security 290
   3. Privacy 290
   4. Bias and Fairness 290
   5. Transparency 291
   6. Misuse and Harms 291
   7. Compliance 291
  Strategic Framework for Risk Quantification for AI systems 292
   Minimal Risk AI Systems 292
   Transparency Risk AI Systems 293
   High-Risk AI Systems 293
   Unacceptable Risk AI Systems 293
  Implementing Effective AI Risk Management 294
   1. Establish a Risk Management Culture 294
   2. Integrate Risk Management into AI Lifecycle 294
   3. Foster Cross-Functional Collaboration 294
   4. Continuous Monitoring and Improvement 295
  Summary 295

xxiii

TABLE OF CONTENTS

Chapter 24: Leveraging Generative AI: Strategies, Implementation, and Impact 297

Creating Your Value Hypothesis 298
    1. Strategic Assessment 298
    2. Benchmarking Potential Value 299
    3. Short-Term vs. Long-Term Value 299

Prioritizing Key Use Cases 299
    1. Identifying High-Impact Use Cases 299
    2. Industry-Specific Applications 299
    3. Assessing Value and Feasibility 300

Scaling Through Patterns 300
    1. Model Refinement 300
    2. Leveraging Patterns 300
    3. Value from Net-New Creation and Augmentation 300

Selecting Foundational Generative AI Tools 301
    1. Evaluating Technologies 301
    2. Customizing Models 301
    3. Avoiding Tech Debt 301

Defining Solutions to Maximize Value 301
    1. Proprietary Data Integration 302
    2. Lateral Thinking and Patterns 302
    3. Incremental Solutions 302

Assessing Costs and Carbon Impact 302
    1. Comprehensive Cost Assessment 302
    2. Environmental Impact 303
    3. Reputational Considerations 303

Developing, Testing, and Learning 303
    1. Controlled Deployments 303

TABLE OF CONTENTS

    2. Iterative Learning 303

    3. Reevaluating Risks and Governance 304

Scaling and Adaptation 304

    1. Adaptive Scaling 304

    2. Broadening Applications 304

    3. Institutional Knowledge 304

Seizing the Generative AI Opportunity 305

    Checklist for Leveraging Generative AI: Strategies, Implementation, and Impact 305

Summary 313

Chapter 25: Evaluating Generative AI Use Cases: A Comprehensive Framework 315

Key Considerations in Assessing Generative AI Use Cases 315

    1. Margin (Revenue and Cost) 316

    2. Business Model Disruption 317

    3. Operating Model Disruption 317

    4. Competitive Disruption 318

    5. Model Feasibility 318

    6. Drivers of Change 319

    7. Responsible AI 319

Case Study in Generative AI Deployment: A Global Beverage Company 320

    1. Initial Focus on Predictive Maintenance 320

    2. Scaling to Logistics Management 320

    3. Expanding to Precision Agriculture 320

    4. Continuous Learning and Adaptation 321

Checklist for Evaluating Generative AI Use Cases 321

    Scoring System 321

    Financial Impact and Business Model Disruption 321

TABLE OF CONTENTS

Operating Model and Competitive Disruption 322
Feasibility and Readiness 323
Responsible AI and Risk Management 324
Total Scoring 326
Scoring and Interpretation 326
Threshold for Passing 326

Summary 327

Chapter 26: AI Executive Compensation: Insights from Europe and the United States 329

Organizational Structure and Roles 329
   1. Key Leadership Roles 330
   2. Experience and Background 330
   3. Reporting Structures 330

Compensation Insights 331
   1. Base Salaries and Bonuses 331
   2. Equity Compensation 331
   3. Industry-Specific Compensation 332

Additional Insights from Compensation Snapshot 332
   1. Financial Services Dominance 332
   2. Team Size and Compensation 332
   3. Regional Variations within the United States 333
   4. Gender and Ethnic Disparities 333

Diversity and Inclusion 333
   1. Gender Diversity 333
   2. Ethnic Diversity 334

Key Considerations for AI Leadership 334
   1. Competitive Compensation Packages 334
   2. Strategic Reporting Lines 335

3. Fostering Diversity and Inclusion 335
　　4. Continuous Professional Development 335
　Summary.. 340

Chapter 27: Strategic Insights on the Reporting Structures of AI Executives 341
　Broad Patterns in Reporting Structures 342
　Strategic Directions for Executives and Boards 343
　　Align Reporting Structures with Strategic Objectives 343
　　Foster a Collaborative Culture 344
　　Empower AI Executives with Resources and Authority 344
　　Promote Ethical and Responsible AI 345
　Strategic Recommendations 345
　　Checklist for AI Executive Reporting Structures: A Strategic Guide for Boards and C-Suites 347
　　Common Themes for All Reporting Structures 347
　　Specific Reporting Structures 348
　　Strategic Recommendations 349
　　Total Scoring 350
　　Scoring and Interpretation 350
　　Threshold for Passing 351
　Summary.. 351

Chapter 28: Governance and Oversight of AI Systems 353
　Key Governance Strategies 353
　Practical Steps for Boards 354
　Ensuring Ethical AI Practices 354
　　Ethical Guidelines for AI 355
　　Monitoring and Evaluation 355

TABLE OF CONTENTS

- Understanding AI's Role in Corporate Strategy 356
  - Strategic Opportunities of AI 356
  - Board's Oversight Role 356
- Establishing Effective AI Governance 357
  - Governance Framework 358
  - Evaluating AI Projects 358
- Practical Steps for AI Oversight 359
  - Proactive Oversight Actions 359
  - Mitigating Directors' and Officers' Liability 360
- Leveraging AI for Business Value 360
  - Business Value Applications 360
- Strategic Considerations for Boards 361
  - Key Board Responsibilities 361
- Checklist for Governance and Oversight of AI Systems 361
  - Scoring System 361
  - Governance and Oversight 362
  - Practical Steps for Boards 362
  - Ensuring Ethical AI Practices 363
  - Monitoring and Evaluation 364
  - Understanding AI's Role in Corporate Strategy 364
  - Establishing Effective AI Governance 365
  - Evaluating AI Projects 366
  - Proactive Oversight Actions 367
  - Mitigating Directors' and Officers' Liability 368
  - Total Scoring 369
  - Scoring and Interpretation 369
  - Threshold for Passing 369
- Summary 369

TABLE OF CONTENTS

Chapter 29: Assessing and Advancing AI Maturity in Organizations 371

   Stages of AI Maturity..371

      AI Unaware 372

      AI Aware 373

      AI Ready..373

      AI Competent..374

      From AI Unaware to AI Aware 375

      From AI Aware to AI Ready...375

      From AI Ready to AI Competent 376

      Key Responsibilities of Boards 376

   Summary..377

Index..379

About the Author

Rohan Sharma is a business executive and philanthropic leader with a strong network in industry, academia, and investment community and a consistent track record of driving revenue growth and operational performance improvement. Rohan's expertise also includes leading initiatives in automation, data science, and AI for top Fortune 100 companies and in Hollywood, including Apple, Disney, AT&T, and Honda.

Rohan Sharma is an award-winning technology executive and TEDx speaker who has led AI products and digital transformations at Apple, Disney, and Fortune 100s. He is a Stanford Seed strategy consultant and advisory board member at Frost & Sullivan and Contact Center Expo. Currently a Consumer Data & Analytics leader at Mattel, he serves as an Industry Fellow at IE Business School and strategic advisor to UCLA Anderson school of Management. He also mentors at TechStars and UC San Diego School of Management and serves on HBR advisory council and Forbes business council. His book AI & Boardroom (Springer Nature) captures proven frameworks for enterprise AI adoption. A USC-trained engineering leader and author of "Minds of Machines," Rohan speaks globally on AI transformation, digital leadership, customer experience and Analytics at prestigious C-suite forums.

Acknowledgments

I would like to express my deepest gratitude to my wife Ekavali, whose unwavering support, love, and understanding have been the driving force behind the completion of this book. Your belief in me and your encouragement during challenging times have been invaluable.

I am also immensely grateful to my parents for their endless encouragement, sacrifices, and belief in my abilities. Your unwavering support and guidance have shaped me into the person I am today, and I am forever thankful for your love.

I would like to extend my heartfelt thanks to my amazing friends who have been with me throughout this journey. Your encouragement, constructive feedback, and willingness to lend a helping hand have been instrumental in shaping the ideas presented in this book. Not to mention being thankful for your understanding in bailing out on our gatherings.

I would also like to express my appreciation to the countless individuals, mentors, and colleagues who have contributed to my understanding and knowledge in the field of business. Your insights, discussions, and shared experiences have enriched my perspective and influenced the content of this book.

Lastly, I am grateful to the readers and supporters of this book. Your interest, feedback, and engagement inspire me to continue sharing my ideas and insights.

Without the support and love of these incredible individuals, this book would not have been possible. Thank you all for being an integral part of my life and for contributing to the realization of this project.

With deepest gratitude, Rohan Sharma.

CHAPTER 1

Introduction

Companies are now going all-in on data and artificial intelligence (AI). They're playing around with fancy names like Generative AI, LLMs, vector databases, advanced analytics, data science, and machine learning, but their mission remains as clear as ever: make more money and work smarter, not harder. Organizations as always are on a quest to boost their revenue, streamline operations, cut costs, and cook up some tantalizing data-powered offerings that will leave their competitors drooling and stock markets rewarding them with runaway valuations.

It's common knowledge that in this digital era, to stay ahead of the game, businesses need to have an agile operating model, have a digital footprint, and, most importantly, be smart about capital allocation. And where do they find this magical smartness on where to play, how to win and how to allocate appropriately? Well, it depends on when and who you ask, but most will say data and AI in general and generative AI to be particular. Both digitally native and legacy organizations are feeding their internal processes and products with brain-boosting data and machine learning model juice, just so that they can raise their hand in their earnings calls and annual conferences and boast out "We too have a Generative AI play."

Boards and C-suites of any organization that intends to stay alive in the next 10 years have reinvigorated focus on data and AI initiatives with high hopes of transforming their businesses and attracting top talent. However, if history is any indicator, after a couple of years, many of these programs are exhibiting signs of weariness and unfulfilled expectations. Boards and

CHAPTER 1   INTRODUCTION

executive leaders typically end up being deeply dissatisfied with the pace of progress. Any guesses why? Well, to start with, Data and AI are still niche activities in their businesses, not a core competence of the business, and that is where the problem begins.

The reality is that there are no shortcuts to become a data driven, AI turbocharged organization. Amazon (Alexa), Google, Microsoft (OpenAI), and Facebook (image recognition), employed distinct business strategies to achieve their current AI market dominance and by that extension, trillion-dollar valuations. However, their shared triumph can be attributed to their visionary recognition of the significance of data and their early positioning in this realm. All of them adopted internal approaches, prioritizing the continuous development of human capabilities while simultaneously focusing on the internal development, experimentation, and implementation of cutting-edge technologies albeit in different ways. And that's precisely the reason why they are all extremely well positioned to monetize the coming AI payday.

For traditional brick-and-mortar companies that have yet to fully embrace digital transformation, it will be an uphill battle. These companies often have established bureaucratic processes, employees with limited digital proficiency, and legacy infrastructure. Overcoming these obstacles requires unwavering determination and persistence from the company's leadership and sadly sometimes even that is not enough.

Imagine this: a unionized 80-year-old brick-and-mortar auto manufacturer (take GM for example), trying to integrate data and AI into all aspects of decision-making, from strategic planning to day-to-day operations, supported by key performance indicators that prioritize data-driven insights. Their leadership meetings, ranging from the board to the C-suite and senior management, have discussions revolving around developing data and AI capabilities.

> *Do you see that happening? Long shot in my opinion.*
> *Maybe some for optics.*

CHAPTER 1 INTRODUCTION

To really get to speed with the rate of AI transformation happening, C-suites and boards at most digitally non-native, legacy organizations need to be highly involved in all aspects of execution of data and AI strategies and the capabilities that drive their unique business strategies. Without a 100% commitment from the top leadership, digital and AI adoption will be patchy and unsustainable at best and complete digital transformation is a tall order.

In some cases, leadership recognizes the significance of adopting a data- and AI-driven approach but feel limited in their own knowledge on where to start and the operating model to adopt. One effective approach is to incorporate a customized data and AI workshop as part of the leadership strategy sessions. It's crucial to understand though that some business leaders mistakenly focus on statistics, computer science, and coding when seeking to enhance their understanding of AI. While coding is a crucial skill for data scientists and engineers, business leaders are better served by focusing their efforts on creating an enabling environment for data and AI within the company. This includes setting clear business objectives, laying out the AI strategy, defining the right operating model for AI to operate at scale, understanding the AI privacy, AI regulations and even AI copyright laws to commit to necessary investments, preventing any legal troubles for the organization, and establishing an effective operating model and organizational structure for data and AI.

During the course of this book, we will explore precisely the topics that executives and boards will need a thorough understanding of to make informed capital allocation and strategic decisions for their organizations.

CHAPTER 1 INTRODUCTION

Summary

The divide between digital natives and traditional enterprises looms large. Can legacy organizations truly transform into data-driven, AI-powered entities, or will they be left in the digital dust? The path forward demands more than mere lip service to AI initiatives – it requires unwavering commitment from the C-suite and board, a fundamental reimagining of business processes, and a willingness to invest in both technology and human capital. For executives, the imperative is clear: focus not on the intricacies of coding, but on crafting an environment where AI can thrive, aligning it with clear business objectives and navigating the complex landscape of AI strategy, privacy, and regulation. The AI train is leaving the station – will your organization be on board, or watching from the platform?

CHAPTER 2

AI Governance

Introduction

AI Governance is the essential set of frameworks, standards, processes, and tools to provide right guardrails to implementing AI technologies in organizations, government, and rolling out to society at large responsibly.

They key goal in laying out AI Governance standards is to bring humans into the loop at the right time and bridge the gap between the potential of AI and the humans that are building, monitoring, controlling, explaining, and reporting on the AI systems.

Technology is changing at a much faster rate than humans can adapt to it. AI Governance is needed so that we do not end up in the similar spot we ended up with social media wave which was not a pretty story. Systemic bias within some social media organizations? Private user data sold to advertisers? Numerous data breaches? We all have seen this move before, the problem is without right guardrails and AI governance it will be orders of magnitude worse.

As Sam Altman recently said in a recent Senate Congressional hearing "If AI goes wrong, it can go very wrong."

CHAPTER 2   AI GOVERNANCE

Drivers for AI Governance

We discussed above "Why we need AI Governance?" Let's discuss "Why we need it NOW?"

Driver 1: Legal and Regulatory Compliance

Most governments in the developed world are acting on AI regulations, and they are acting fast. AI Governance is imperative to complying with existing and new laws being introduced across the United States, Canada, and the EU. In many cases, these laws and regulations differ in their scope, compliance standards, and by extension, penalties for non-compliance.

EU for instance has some of the most stringent GDPR Data Privacy laws, and penalties for non-compliance can be pretty significant. Similarly, their proposed Artificial Intelligence Act (AIA 2023) is considered the most comprehensive and strict regulation that will be in place.

The United States, Singapore, and China, on the other hand, have guidelines. Guidelines are not regulations. They do not have enforcement teeth, with civil or criminal penalties.

A right governance model is therefore extremely crucial to ensure organizations are operating within the legal and regulatory framework in the jurisdiction that they are operating. A lapse could result in, well just a $6.5 billion fine Facebook was slapped with in the EU in 2023.

Driver 2: Ensuring Accountability

AI is a rapidly progressing technology and most organizations do not have clarity on who clearly owns the agenda for responsible development and adoption of AI. Well, to be honest, that's the case with most of their legacy technology as well because what is an organization without power struggles?

Coming back to who owns the Data and AI agenda? It could be the CIO, but how would business unit leaders experiment and adopt new tools and platforms? How does an organization's legal team make sure the AI tools being adopted comply with legal regulation and data privacy policies? How does a company's HR leadership make sure the AI tools and platforms being developed and adopted do not have inherent bias? How does an organization's marketing leadership ensure that all external communications generated by AI tools and platforms maintain and reinforce brand promise?

A well-thought-out governance model is therefore a starting point to laying out right accountabilities.

AI Governance: Approaches for Federal and State Governments

Just like organizations, governments need to lay out AI governance standards.

There are several approaches which have been discussed over the years, approaches from testimonies at various congressional hearings to approaches proposed by think tanks like Brookings Institute.

One approach is to create new laws and regulations at the federal level to govern AI, much like the European Union has done with GDPR and is now attempting to do with AIA.

Another approach being discussed is to develop standards and guidelines for the development and use of AI much like Singapore and China have done at the federal level. As an example, Singapore's Model AI Governance Framework sets a global standard, providing detailed guidance to private sector organizations on ethical and governance issues in AI deployment. Emphasizing explainability, transparency, and fairness, these guidelines aim to foster public understanding and trust in technology. Singapore's visionary approach serves as a reference point, illuminating the path for responsible and beneficial AI implementation worldwide.

The United States does have some guidelines in place through the National Institute of Standards and Technology (NIST), Federal Trade Commission, and the White House's AI Bill of Rights, but a lot needs to be done in this space.

A third approach is to create independent oversight bodies, to monitor the development and use of AI. It could be empowering existing agencies like National Institute of Standards and Technology (NIST) in the United States with more enforcement power or create entirely new agencies for enforcement of laws and regulations pertaining to AI.

The best approach to AI governance in the United States will likely involve a combination of these different approaches for uniform adoption of responsible AI across a spectrum of jurisdictions.

AI Governance: Approach for Organizations

Holistic AI governance drives accountability, transparency, and alignment with business strategy. It also establishes standards, methodologies, and oversight, enabling technical teams to focus on performance metrics and executing a roadmap.

A well-laid-out AI governance framework even facilitates build vs. buy decisions and compliance with regulations, with dedicated governance teams managing risk and facilitating audits. For example, AI Governance teams can be instrumental in shortlisting vendors which comply with organization's data security, data privacy and compliance policies.

Finally, effective AI governance framework complements technical AI expertise and strengthens enterprise capabilities, ensuring scalability, risk mitigation, and sustained ROI on AI investments, because what is windfall in revenue worth if it brings in a wave of lawsuits against the organization? A limited-risk approach for organizations is to create a cross-functional leadership team with complete buy-in from top leadership to lay out the AI governance framework, develop guardrails around developing and adopting AI responsibly and laying out change management and communication plan to bring transparency around decisions being taken to the rest of the organization.

Illustrates where AI governance layer sits in the organization and how it can be the glue between executive leadership and technical AI teams executing on AI roadmap.

An Integrated Framework for AI Deployment and Governance

Successfully deploying AI at scale requires a well-structured approach that encompasses all stages of the AI lifecycle – from discovery to operation. This chapter provides a comprehensive framework for AI deployment and governance, integrating key activities across data collection, model development, validation, execution, and ongoing operation. By understanding this framework, corporate boards and senior executives can ensure their AI initiatives are effective, ethical, and aligned with strategic business objectives.

CHAPTER 2   AI GOVERNANCE

The AI Deployment and Governance Framework

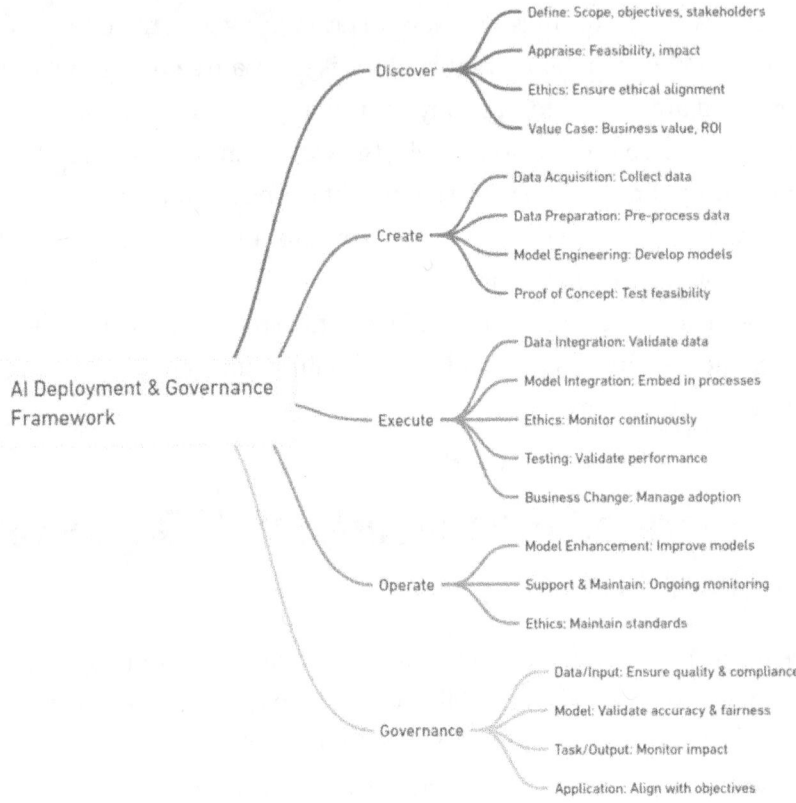

The framework above outlines the key stages and activities involved in deploying AI at scale. This approach ensures that AI initiatives are systematically managed, from initial discovery through to ongoing operation and improvement.

1. Discover

The discovery phase involves defining the AI project, appraising its potential, ensuring ethical considerations, and establishing a clear value case. This phase sets the foundation for successful AI deployment by aligning the project with business goals and ethical standards.

Key Activities

- **Define**: Identify the AI project's scope, objectives, and stakeholders.

- **Appraise**: Assess the feasibility and potential impact of the AI project.

- **Ethics**: Consider ethical implications and ensure adherence to ethical guidelines.

- **Value Case**: Establish the business value and expected return on investment.

2. Create

The creation phase focuses on data acquisition, preparation, and model development. This phase includes rigorous data processing and proof-of-concept development to ensure that AI models are robust and aligned with business needs.

Key Activities

- **Data Acquisition**: Collect and ingest data from various sources.

- **Data Preparation**: Pre-process and sample data to ensure quality and relevance.

- **Model Engineering**: Develop and refine AI models, incorporating ethical considerations.
- **Proof of Concept**: Test the model's feasibility and business value.

3. Execute

The execution phase involves integrating and validating AI models, ensuring they are technically sound and aligned with business processes. This phase includes comprehensive testing and change management to facilitate smooth deployment.

Key Activities

- **Data Integration/Validation**: Ensure data is correctly integrated and validated for AI models.
- **Model Integration**: Embed AI models into business processes and systems.
- **Ethics**: Continuously monitor ethical considerations during integration.
- **Technical Strategy (RTL)**: Implement the technical roadmap for AI deployment.
- **Testing Lifecycle**: Conduct thorough testing to validate model performance and reliability.
- **Business Change**: Manage organizational changes to support AI adoption.

4. Operate

The operation phase ensures that AI systems are maintained, monitored, and continuously improved. This phase focuses on ongoing support, model enhancement, and adherence to ethical standards.

Key Activities

- **Model Enhancement**: Continuously improve AI models based on feedback and performance metrics.

- **Support and Maintain**: Provide ongoing support and monitoring to ensure AI systems operate effectively.

- **Ethics**: Maintain ethical standards and address any emerging ethical issues.

AI Lifecycle Governance

Governance is a critical component that spans across all stages of the AI lifecycle. It ensures that AI initiatives are aligned with regulatory requirements, ethical standards, and organizational policies. Effective governance involves a structured approach to monitoring, evaluating, and improving AI systems.

Governance Framework

- **Data and Input Governance**: Ensuring data quality, security, and compliance during collection and processing.

- **AI Model Governance**: Validating and verifying AI models to ensure accuracy, fairness, and transparency.

- **Task and Output Governance**: Monitoring AI outputs and their impact on business processes and stakeholders.
- **Application Context Governance**: Ensuring AI applications are aligned with business objectives and ethical guidelines.

Strategic Implementation of AI Deployment and Governance

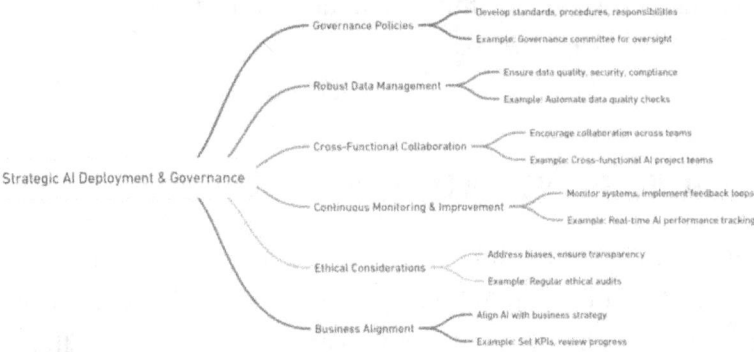

To implement an effective AI deployment and governance framework, organizations should follow these strategic steps:

1. Establish Clear Governance Policies

Develop and implement governance policies that outline standards, procedures, and responsibilities for AI initiatives. This ensures consistency and compliance across all stages of the AI lifecycle.

Example: Create a governance committee to oversee AI projects and ensure adherence to ethical guidelines and regulatory requirements.

2. Invest in Robust Data Management

Ensure robust data management practices to maintain data quality, security, and compliance. This includes implementing data validation protocols, access controls, and encryption measures.

Example: Use advanced data management tools to automate data quality checks and ensure compliance with data protection regulations.

3. Foster Cross-Functional Collaboration

Encourage collaboration between data providers, IT operations, data science teams, and business units to ensure AI initiatives are well-integrated and aligned with business goals.

Example: Establish cross-functional teams for AI projects, including members from different departments to provide diverse perspectives and expertise.

4. Implement Continuous Monitoring and Improvement

Continuously monitor AI systems and implement feedback loops to improve model performance and address any issues. This involves regular audits, performance reviews, and updates to AI models.

Example: Use AI monitoring tools to track model performance in real time and identify areas for improvement.

5. Prioritize Ethical Considerations

Ensure ethical considerations are prioritized throughout the AI lifecycle. This includes addressing biases, ensuring transparency, and protecting user privacy.

CHAPTER 2   AI GOVERNANCE

Example: Conduct regular ethical audits to identify and mitigate biases in AI models and ensure compliance with ethical guidelines.

6. Align AI Initiatives with Business Strategy

Align AI initiatives with the organization's broader business strategy to ensure they drive value and support strategic objectives. This involves setting clear goals, KPIs, and metrics for AI projects.

Example: Define strategic KPIs for AI projects, such as increased efficiency, cost savings, and enhanced customer experiences, and regularly review progress against these metrics.

Crafting a Robust Operational AI Governance Framework

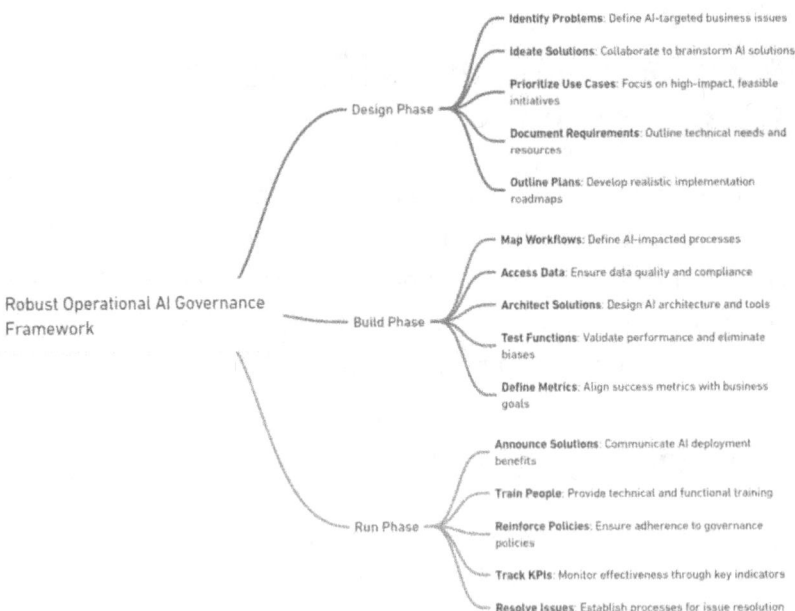

Design Phase: Laying the Foundation

The first step in creating a successful AI governance framework is the design phase. This stage focuses on identifying and defining the problems AI aims to solve, ideating potential solutions, and outlining the necessary plans and requirements.

1. **Identify Problems**: Clearly articulate the business problems that AI can address. Engage with stakeholders across the organization to understand pain points and areas where AI can add value.

2. **Ideate Solutions**: Brainstorm and develop AI solutions that align with the identified problems. This involves collaboration between business units and AI experts to generate innovative ideas.

3. **Prioritize Use Cases**: Not all AI initiatives can be pursued simultaneously. Prioritize use cases based on their potential impact, feasibility, and alignment with strategic goals.

4. **Document Requirements**: Create detailed documentation outlining the requirements for each prioritized use case. This includes technical specifications, data needs, and resource allocation.

5. **Outline Plans**: Develop comprehensive plans that cover the implementation roadmap, timelines, and milestones. Ensure that the plans are realistic and aligned with organizational capabilities.

CHAPTER 2   AI GOVERNANCE

Build Phase: Constructing the Solution

Once the design phase is complete, the focus shifts to building the AI solutions. This phase involves mapping workflows, accessing and preparing data, and developing and testing AI models.

1. **Map Workflows**: Define the workflows and processes that will be impacted by AI solutions. This helps in understanding the integration points and ensuring smooth implementation.

2. **Access Data**: Secure access to the necessary data for training and deploying AI models. Ensure data quality and compliance with privacy regulations.

3. **Architect Solutions**: Design the technical architecture of AI solutions. This includes selecting the appropriate algorithms, frameworks, and tools.

4. **Test Functions**: Rigorously test AI models to ensure they meet performance criteria and are free from biases. Use real-world scenarios to validate model effectiveness.

5. **Define Metrics**: Establish clear metrics for measuring the success of AI solutions. These metrics should align with business objectives and provide insights into the performance and impact of AI initiatives.

Run Phase: Operationalizing AI

The final phase of the AI governance framework is the run phase, where AI solutions are deployed and managed in a production environment. This stage focuses on operationalizing AI, training personnel, and continuously monitoring performance.

1. **Announce Solutions**: Communicate the deployment of AI solutions across the organization. Highlight the benefits and changes to existing workflows to ensure a smooth transition.

2. **Train People**: Provide comprehensive training to employees on how to interact with and leverage AI solutions. This includes both technical training for AI specialists and functional training for end users.

3. **Reinforce Policies**: Ensure that AI governance policies are reinforced and adhered to. This includes data governance, ethical guidelines, and compliance with relevant regulations.

4. **Track KPIs**: Monitor key performance indicators (KPIs) to evaluate the effectiveness of AI solutions. Use these insights to make data-driven decisions and continuous improvements.

5. **Resolve Issues**: Establish a process for promptly identifying and resolving any issues that arise during the operational phase. This includes technical glitches, data quality problems, and user feedback.

CHAPTER 2   AI GOVERNANCE

AI Governance Checklist for Board and C-Suite Executives

To ensure robust AI governance, board members and C-suite executives must ask the right questions to guide their organizations in navigating the complexities of AI implementation. This checklist incorporates a scoring system to quantitatively assess the strength of your AI governance framework.

Scoring System

- **0 points:** Not addressed
- **1 point:** Partially addressed
- **2 points:** Fully addressed, but needs improvement
- **3 points:** Fully addressed and well-executed

Environmental Layer: Setting the Stage for AI Governance

1. **Regulatory Compliance**
    - Are we compliant with all relevant hard laws and regulations governing AI use?
        - Score: 0-3
    - Have we identified and documented the legal requirements specific to our industry and AI applications?
        - Score: 0-3

2. **Principles and Guidelines**
   - Do we have established principles and guidelines that govern the ethical use of AI within our organization?
     - Score: 0-3
   - How are these guidelines communicated to and enforced across all levels of the organization?
     - Score: 0-3
3. **Stakeholder Engagement**
   - How do we engage with stakeholders to understand their concerns and expectations regarding our use of AI?
     - Score: 0-3
   - Are we transparent about our AI practices and responsive to stakeholder feedback?
     - Score: 0-3

Organizational Layer: Aligning AI with Business Strategy

1. **Strategic Alignment**
   - How does our AI strategy align with our overall business strategy and objectives?
     - Score: 0-3
   - Are we leveraging AI to create strategic value and competitive advantage?
     - Score: 0-3

2. **Value Alignment**
   - How do our AI initiatives align with our organizational values and mission?
     - **Score: 0-3**
   - Are we using AI to enhance our commitment to corporate social responsibility?
     - **Score: 0-3**
3. **Communication and Engagement**
   - How do we ensure continuous communication and engagement about AI initiatives within the organization?
     - **Score: 0-3**
   - Are there clear channels for feedback and knowledge-sharing among teams working on AI projects?
     - **Score: 0-3**

AI System Layer: Ensuring Operational Excellence

1. **AI System Design and Operations**
   - Are our AI systems designed with robustness and reliability in mind?
     - **Score: 0-3**
   - What processes do we have in place to ensure the operational integrity of our AI systems?
     - **Score: 0-3**

2. **Algorithm Design and Operations**
   - How do we ensure our AI algorithms are fair, unbiased, and transparent?
     - **Score: 0-3**
   - What measures are in place to regularly audit and improve algorithm performance?
     - **Score: 0-3**
3. **Risk and Impact Management**
   - How do we assess and mitigate the risks associated with AI deployment?
     - **Score: 0-3**
   - What frameworks are in place for continuous risk management and impact assessment?
     - **Score: 0-3**
4. **Data Operations**
   - How do we ensure the quality, security, and privacy of the data used in our AI systems?
     - **Score: 0-3**
   - Are our data management practices aligned with industry standards and best practices?
     - **Score: 0-3**
5. **Development Operations**
   - Do we have a structured process for the development and deployment of AI models?
     - **Score: 0-3**

CHAPTER 2   AI GOVERNANCE

- How do we integrate ethical considerations into our AI development lifecycle?
  - Score: 0-3

Accountability and Compliance

1. **Ownership and Accountability**
   - Who is accountable for the oversight and governance of AI within our organization?
     - Score: 0-3
   - How is accountability for AI outcomes distributed among different stakeholders?
     - Score: 0-3
2. **Transparency and Contestation**
   - How do we ensure transparency in our AI decision-making processes?
     - Score: 0-3
   - What mechanisms do we have in place to address and rectify any issues or biases identified in our AI systems?
     - Score: 0-3
3. **Regulatory Compliance**
   - How do we stay updated on emerging regulations and standards related to AI?
     - Score: 0-3

- Are our compliance measures proactive and adaptable to regulatory changes?
  - **Score: 0-3**

Continuous Improvement and Knowledge Flow

1. **Integration and Knowledge Flows**
   - How do we integrate AI knowledge and best practices across different functions and teams?
     - **Score: 0-3**
   - Are there mechanisms in place to ensure continuous learning and improvement in our AI governance practices?
     - **Score: 0-3**

Total Scoring

- **Total Maximum Score: 90 points**

Interpretation

- **75–90 points:** Excellent AI governance. The organization has a comprehensive and well-executed AI governance framework.
- **50–74 points:** Good AI governance. There are areas for improvement, but the overall framework is solid.

- **25–49 points:** Fair AI governance. Significant improvements are needed to strengthen the AI governance framework.

- **0–24 points:** Poor AI governance. The organization lacks a coherent AI governance framework and needs to take immediate action.

By addressing these questions and scoring the responses, boards and C-suite executives can quantitatively assess their AI governance framework. This proactive approach ensures that AI initiatives are aligned with organizational values, regulatory requirements, and stakeholder expectations, ultimately driving sustainable business success.

Summary

As AI accelerates beyond human adaptability, the imperative for robust governance becomes not just important, but existential. Will organizations and governments act swiftly to implement comprehensive AI governance frameworks, or risk stumbling into a technological minefield blindfolded? The path forward demands a delicate balance: marrying regulatory compliance with ethical considerations, aligning AI initiatives with business strategies, and fostering a culture of accountability and continuous improvement. For boards and C-suite executives, the challenge is clear – transform AI governance from a nebulous concept into a quantifiable, actionable framework that permeates every level of the organization. The AI revolution is here; will your governance be its guiding light or its Achilles' heel?

CHAPTER 3

AI Regulation

As Artificial intelligence (AI) continues to evolve at an exponential pace, what we need to realize is that the need for regulating it is increasing at a runaway pace as well.

AI systems are becoming exponentially powerful, complex, and self-optimizing. They are starting to have transformative impact on people's lives and civilization as we know it.

So, without a doubt, the time to lay out regulations is now. As it was said in a recent congressional testimony by Sam Altman, CEO of OpenAI. "If this technology goes wrong, it can go very wrong."

CHAPTER 3   AI REGULATION

Approaches to AI Regulation

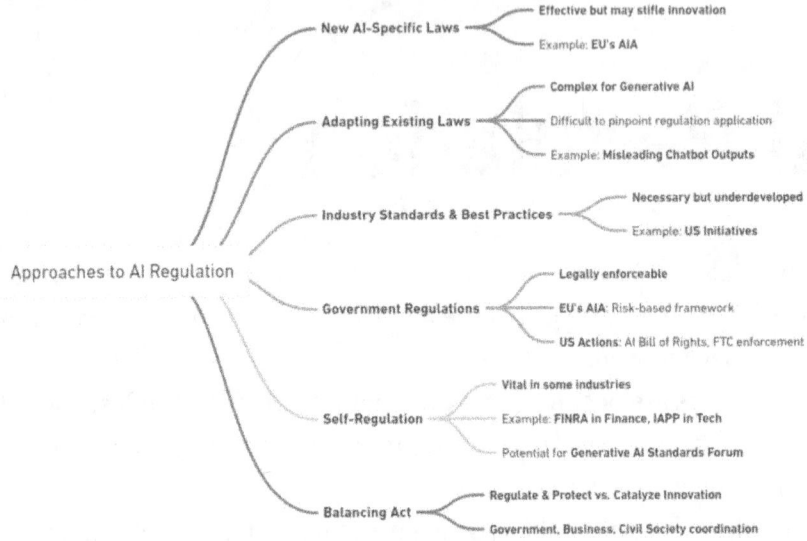

There are several different approaches being considered with respect to AI regulation. One approach is to create new laws and regulations specifically for AI, similar to what the EU is approaching with AIA. It's an effective approach, but can be stifling to innovation in the long run.

Another approach is to adapt existing laws and regulations to apply to AI systems, an approach which has been debated extensively in Congress but is easier said than done because the Generative AI system is so complex that it's hard to interpret existing laws and regulations on them. For example, in case of Generative AI, it's not readily apparent to single out where the regulation might be applied. Let's consider a user-generated script that runs on plugins that connect to fine-tuned models that connect to foundational LLM models that connect to datasets. In the above scenario, where and how does a regulatory and compliance policy of misinformation in a chatbot get applied?

Do you regulate the use of Large Language Models like GPT4? The datasets they are trained on, that is, the Internet? The cloud hosting providers like AWS, Azure, GCP? The GPU providers like Nvidia? The product manufacturers? Or the end-user application creators like AutoGPT.

Although it's not abundantly clear when, how, and where compliance liability rests, it's important to start the conversation.

A third approach that's being discussed is to develop industry standards and best practices for the development and use of AI, some of which has been done in the United States, but a lot remains to be done.

The best approach to AI regulation is likely to involve a combination of these different approaches. It is important to find a way to balance the need to protect people from the risks of AI with the need to allow AI-enabled innovation to blossom which will have transformative impacts on some of the biggest problems of our time.

It's important to note that regulations can take many forms, although we primarily relate it to government regulations. But there is also self-regulation. Let's look at the current state of AI regulation as it relates to both types.

1. Government regulations
2. Self-regulation

Government Regulations

These are laws put in place to regulate the development and usage of a particular technology. Laws and government regulations have teeth, that is, there are civil and criminal penalties for non-compliance.

CHAPTER 3   AI REGULATION

European Union

The European Union is further along in the laying of laws and regulations pertaining to AI. The EU has finalized European Union: Artificial Intelligence Act (AIA).

The AIA is a proposed regulation that would govern the development and use of AI in the European Union. The regulation would create a risk-based framework for regulating AI, with different requirements for AI systems that pose different levels of risk. The AIA would also establish a European Artificial Intelligence Board to oversee the implementation of the regulation.

The United States

In the United States, the Federal government via the Commerce Department has started to take steps to create AI safety rules.

The White House have also released AI Bill of Rights, which is a non-binding document that outlines five key principles for the responsible use of artificial intelligence (AI).

Some Federal agencies have also acted against unethical and illegal use of AI. For example, the FTC has taken several enforcement actions against companies that have used AI in a deceptive or unfair manner. In 2021, the FTC fined Facebook $5 billion for using facial recognition technology in a deceptive manner.

Similarly, there is guidance from Office of Management and Budget (OMB) on Regulation of Artificial Intelligence Applications. This guidance, issued by OMB in November 2020, provides guidance to federal agencies on how to regulate AI applications. The guidance emphasizes the need for a risk-based approach to regulation, and it encourages agencies to consider the potential benefits of AI, as well as the potential risk. On similar lines, The National Institute for Standards and Technology put out an AI Risk Management Framework.

Role of Government in Regulating Technology

This brings us to the question, what should be the role of the government in regulating AI and similar technologies with massive impact on society?

The image below taken from European Commission's OECD website accurately lays out the scope of government's involvement. It concludes that the role of government should be that of a regulator and protector on the one hand, while being a catalyst and enabler on the other. Not the easiest position to be in and hence it will take a few iterations for regulations to find their own balance.

A Self-Regulatory Framework

Another kind of regulation, albeit less common, is self-regulation which plays a vital role in various industries. For instance, in finance, bodies like FINRA establish rules for securities firms and brokers. In the UK, Healthcare professions have regulatory bodies like the General Medical Council, ensuring ethical conduct and education standards. Even, advertising industry has the Advertising Standards Authority, keeping ads legal. Technology, too, has self-regulatory bodies like the International Association of Privacy Professionals, safeguarding privacy, and cybersecurity.

On similar lines, we could very well have a Generative AI standards forum that unites top AI leaders and the broader community to regulate AI technology and products effectively and responsibly.

CHAPTER 3   AI REGULATION

The future of AI regulation is uncertain. However, one thing is for sure: AI is a transformative technology that will have an everlasting impact on our society and time is now to start thinking about a robust regulatory framework around AI which lays out foundational guardrails to prevent misuse, outreach of these powerful technologies by bad actors, including state actors, but are not too stifling that America's competitiveness and innovation are jeopardized. And for that balance to happen, active involvement, careful planning, and coordination between governments, businesses, and civil society is imperative.

Checklist for AI Regulations Considerations for Board and C-Suite

Scoring System

- **0 points:** Not addressed
- **1 point:** Partially addressed
- **2 points:** Fully addressed, but needs improvement
- **3 points:** Fully addressed and well-executed

General AI Regulatory Compliance

1. **Awareness of Existing Regulations** (Score: 0-3)
   - Are we fully aware of current AI regulations in our operational jurisdictions?
   - Have we identified which of our AI systems are affected by these regulations?

2. **Government Regulations and Compliance** (Score: 0-3)

   - Are we in compliance with existing government regulations, such as the EU's Artificial Intelligence Act (AIA) or US federal guidelines?
   - Do we have a process for staying updated with changes in AI regulations globally?

3. **Risk-Based Framework** (Score: 0-3)

   - Have we implemented a risk-based framework to evaluate and categorize our AI systems according to their potential risk levels?
   - Is this framework aligned with the regulatory guidelines provided by governing bodies?

4. **Regulatory Reporting and Documentation** (Score: 0-3)

   - Do we have a robust system for documenting our AI systems' compliance with relevant regulations?
   - Are we prepared for potential audits or inquiries from regulatory authorities?

Legal and Ethical Considerations

5. **Legal Liability and Accountability** (Score: 0-3)

   - Have we clearly defined legal liability and accountability for our AI systems?
   - Are our contracts and agreements with third parties updated to reflect AI-related liabilities?

6. **Ethical AI Use** (Score: 0-3)

    - Have we established ethical guidelines for the use of AI within our organization?
    - Do these guidelines align with broader societal values and regulatory expectations?

Self-Regulation and Industry Standards

7. **Self-Regulatory Framework** (Score: 0-3)

    - Do we participate in or adhere to any industry self-regulation standards or frameworks?
    - Are we contributing to the development of industry standards for AI regulation?

8. **Best Practices Implementation** (Score: 0-3)

    - Have we implemented best practices for AI development, deployment, and monitoring within our organization?
    - Do these practices ensure transparency, fairness, and accountability in AI systems?

9. **Internal Audits and Reviews** (Score: 0-3)

    - Are we conducting regular internal audits and reviews of our AI systems to ensure compliance with both legal and self-regulatory standards?
    - Are there mechanisms in place to address and rectify non-compliance issues?

Future-Proofing and Innovation

10. **Adapting to Future Regulations** (Score: 0-3)

    - Do we have a strategic plan for adapting to potential future AI regulations?

    - Is there an ongoing dialogue with policymakers to influence and prepare for regulatory changes?

11. **Balancing Regulation and Innovation** (Score: 0-3)

    - Are we balancing regulatory compliance with the need to innovate and remain competitive?

    - Are we avoiding over-regulation that could stifle innovation while ensuring safe and ethical AI use?

12. **Training and Awareness Programs** (Score: 0-3)

    - Are we conducting regular training and awareness programs for our employees on AI regulations and compliance requirements?

    - Are key stakeholders, including board members and C-suite executives, adequately informed about AI regulatory risks and obligations?

Total Scoring

Each question is scored from 0 to 3, with a maximum possible score of 36 across all 12 questions.

CHAPTER 3   AI REGULATION

Scoring and Interpretation

- **0–12**: High risk – Immediate action is required to address significant gaps in AI regulatory compliance.

- **13–24**: Moderate risk – Areas for improvement identified, with a structured plan needed to enhance compliance and readiness.

- **25–36**: Low risk – Robust AI regulatory strategies in place, with continuous monitoring and minor adjustments recommended.

Threshold for Passing

Organizations should aim for a minimum score of **24** to ensure they have adequately addressed key AI regulatory considerations and are prepared to manage regulatory risks effectively.

Summary

As AI's tentacles reach deeper into the fabric of society, the clarion call for regulation grows louder – but will we answer in time? The regulatory landscape is a complex tapestry of government mandates and industry self-regulation, each thread delicately balancing innovation with ethical constraints. For boards and C-suite executives, the challenge is twofold: navigate the existing regulatory labyrinth while preparing for an uncertain future where AI's capabilities may outpace our ability to govern them. The stakes couldn't be higher – effective regulation could be the difference between AI as humanity's greatest ally or its most formidable adversary. As we stand at this technological crossroads, one question looms large: can we craft a regulatory framework nimble enough to harness AI's potential while robust enough to shield us from its perils?

CHAPTER 4

AI Privacy

Introduction

As we increasingly rely on AI systems to drive innovation and efficiency, the risk of privacy breaches has become a ticking time bomb. With the ability to collect, analyze, and predict human behavior, AI systems pose a significant threat to individual privacy. The question is no longer if, but when the next major data breach will occur. As Marlon Brando aptly put it, "Privacy is not something we are merely entitled to, it's an absolute prerequisite." This chapter explores the critical need for robust AI privacy governance, providing a framework for organizations to safeguard sensitive information while leveraging AI technologies.

One of the biggest and loudest concerns about AI is that it could be used to invade people's privacy. AI systems are increasingly being used to collect, analyze, and further train the models using personal data. This data can be potentially used to track people's movements, monitor their activities, and even predict their behavior, even before they execute a certain action. That is certainly encroaching and even borderline manipulative.

Large Enterprises are not insulated to this phenomenon as well. In a recent incident, Samsung had to issue a companywide policy to restrict the use of open-source LLM models since it was used to get help resolving proprietary source code bugs, fixing software used to gather measurement and yield data, and turning sensitive meeting notes into minutes.

CHAPTER 4   AI PRIVACY

For organizations, one of the preliminary problems with open source LLM model is that the data fed into them is often used to further train the bots, which can easily lead to confidential business information being regurgitated if others ask similar questions.

Secondly, queries executed are also visible to LLM model providers, like OpenAI and Google, who may themselves review the content fed to their language models, further risking the exposure of closely guarded corporate secrets.

Third, these open-source models are not mature and bugs in the system can let users' "conversations" with these models to be viewable by other users. In March 2023, OpenAI admitted that a bug in open-source library redis-py caused bits of people's tête-à-tête with ChatGPT to be viewable by other users. Additionally, there is a big elephant in the room. Open-source AI systems are always at high risk of being hacked, which could lead to the unauthorized disclosure of personal data since these technologies and architectures are relatively new and the security framework to protect them is not fully developed.

Finally, potential for these LLM models to be used to predict people's behavior, which could be used to discriminate against them or to target them with intrusive advertising.

Given these massive data privacy concerns, we must ensure that AI development and deployment respect both privacy rights and ethical considerations.

If regulations are a way to enforce that, we needed them yesterday.

If regulations are to be the enforcement mechanism, they should have been implemented already. However, the United States still lacks a comprehensive data privacy law, largely due to Silicon Valley technology companies' lobbyists preventing these bills from reaching the right desks.

CHAPTER 4   AI PRIVACY

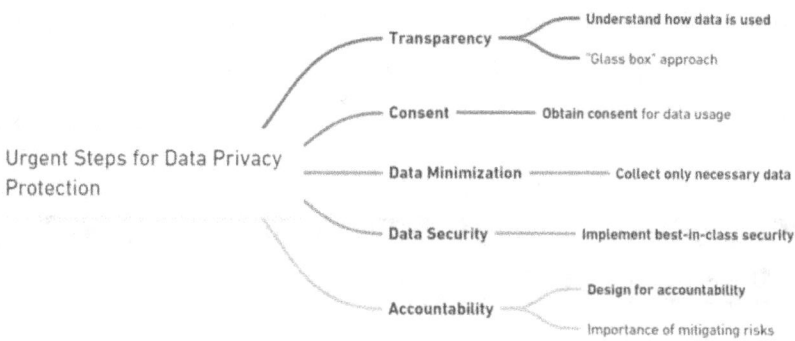

- **Transparency:** People should be given the opportunity to understand how their data is being used by AI systems. "Let it be a glass box rather than a Black Box."

- **Consent:** People should be given the opportunity to consent to the use of their data by AI systems.

- **Data minimization:** AI systems should only collect the data that is necessary for their operation.

- **Data security:** AI systems should have best in class security to protect the privacy of the data that they collect.

- **Accountability:** AI systems should be designed in a way that allows people to hold them accountable for their actions since if this Technology goes wrong, it can go very wrong.

In summary, the United States remains one of the few developed nations without a robust, nationwide data privacy framework. The need for such protection is now more critical than ever, given AI's extraordinary potential for harm if misused by malicious actors.

CHAPTER 4   AI PRIVACY

As it's widely agreed in technology circles, the next data breach will be orders of magnitude bigger than the biggest one that happened, and left you concerned.

Checklist for AI Privacy Considerations for Board and C-Suite

To ensure robust AI privacy governance, board members and C-suite executives must ask the right questions to guide their organizations in navigating the complexities of AI privacy. This checklist incorporates a scoring system to quantitatively assess the strength of your AI privacy framework.

Scoring System

- **0 points:** Not addressed
- **1 point:** Partially addressed
- **2 points:** Fully addressed, but needs improvement
- **3 points:** Fully addressed and well-executed

General Compliance

1. **Lawfulness and Transparency**
    - Are our data processing activities conducted lawfully and fairly?
        - Score: 0-3

- Do we provide clear and transparent information to data subjects about how their data is collected, used, and stored?
  - **Score: 0-3**
2. **Purpose and Storage Limitation**
   - Is data collected for specific, explicit, and legitimate purposes?
     - **Score: 0-3**
   - Do we ensure that data collection is limited to what is necessary for those purposes?
     - **Score: 0-3**
   - Are there protocols in place to ensure data is not stored longer than necessary?
     - **Score: 0-3**

Consent and Data Subject Rights

3. **Consent Management**
   - Do we obtain explicit, informed consent from data subjects before processing their data?
     - **Score: 0-3**
   - Are data subjects aware they can withdraw their consent at any time without detriment?
     - **Score: 0-3**

CHAPTER 4   AI PRIVACY

4. **Rights of Data Subjects**

    - Do we have mechanisms in place to facilitate data subjects' rights to access, rectify, and erase their data?
        - **Score: 0-3**
    - Can data subjects restrict or object to data processing and exercise their right to data portability?
        - **Score: 0-3**

Data Security and Breaches

5. **Data Breach Response**

    - Do we have robust measures to prevent data breaches?
        - **Score: 0-3**
    - Are there clear procedures for detecting, reporting, and managing data breaches to mitigate potential harm to data subjects?
        - **Score: 0-3**

Risk Assessment and Governance

6. **Impact Assessments**

    - Do we conduct regular data protection impact assessments to identify and mitigate risks associated with data processing activities?
        - **Score: 0-3**

7. **Data Protection Officer (DPO)**

    - Have we appointed a Data Protection Officer with the expertise and authority to enforce data protection policies and practices across the organization?

    - Score: 0-3

Privacy by Design and Data Transfers

8. **Privacy by Design**

    - Are data privacy considerations integrated into the design and development of our AI systems from the outset?

    - Score: 0-3

9. **Data Transfers**

    - Do we ensure that data transfers comply with data privacy requirements?

    - Score: 0-3

    - Are appropriate safeguards in place for cross-border data transfers, and do third parties adhere to our data protection standards?

    - Score: 0-3

Organizational Culture and Training

10. **Awareness and Training**
    - Is there a culture of data privacy awareness within our organization?
        - **Score: 0-3**
    - Do we provide regular training to employees on data protection practices, emphasizing the importance of safeguarding personal data and compliance with data privacy policies?
        - **Score: 0-3**

Ongoing Evaluation and Improvement

11. **Continuous Monitoring**
    - Are we continuously monitoring the effectiveness of our data privacy measures?
        - **Score: 0-3**
    - Do we have a system in place for regularly updating our data privacy policies and practices in response to new threats and regulatory changes?
        - **Score: 0-3**

Total Scoring

- **Total Maximum Score: 66 points**

Interpretation

- **55–66 points:** Excellent AI privacy governance. The organization has a comprehensive and well-executed AI privacy framework.

- **40–54 points:** Good AI privacy governance. There are areas for improvement, but the overall framework is solid.

- **25–39 points:** Fair AI privacy governance. Significant improvements are needed to strengthen the AI privacy framework.

- **0–24 points:** Poor AI privacy governance. The organization lacks a coherent AI privacy framework and needs to take immediate action.

Summary

Privacy stands at a critical juncture – will it be sacrificed on the altar of technological progress? The challenges are multifaceted: data-hungry AI systems, vulnerable open-source models, and a lagging regulatory landscape create a perfect storm for privacy breaches. To navigate these turbulent waters, organizations must prioritize transparency, consent, data minimization, and robust security measures, while leveraging tools like the AI Privacy Governance Checklist to fortify their defenses. The next major data breach looms on the horizon, potentially dwarfing all predecessors – will we act now to safeguard privacy, or be left scrambling in the aftermath?

CHAPTER 5

AI Copyright and Intellectual Property

Imagine a world where AI systems are creating new forms of creative expression, such as music, art, and literature. Imagine a world where AI systems are developing new technologies, such as self-driving cars and medical diagnostics. This is the world that we are moving toward. Not so fast, since there is an elephant in the room that has taken backseat to AI Regulation and privacy news headlines.

And this elephant in the room is "Who **owns the Intellectual property of AI-generated content?**"

To address the above question, on March 16, 2023, the Copyright Office initiated an effort to investigate copyright law and policy concerns related to Generative AI.

And they made it abundantly clear

> *If a work's traditional elements of authorship were produced by a machine,* **the work lacks human authorship, and the Office will not register it.**
>
> —Shira Perlmutter, USCO director

CHAPTER 5   AI COPYRIGHT AND INTELLECTUAL PROPERTY

IP and Copyright Laws and How They Apply to AI-Generated Content

IP and Copyright laws are designed to protect innovation and creativity and they do this by giving creators the exclusive right to use their creations for a certain period. This not only allows creators to recoup their investment in creating their works but also encourages them to continue to create new and innovative works.

However, AI poses several challenges to IP law. One of the most significant challenges is that AI can create content that is indistinguishable from human-created content. This makes it extremely difficult to determine who should own the IP rights to AI-generated content.

Another challenge and perhaps a bigger one is that AI can be used to create content that infringes on the IP rights of others since it produces content after training models on proprietary and copyrighted content.

For instance, imagine creating a song with the underlying beats of a well-known artist.

Who owns the right to my new song? Me or machine? Can my machine-generated song be copyrighted by me? Should the original artist, on whose song the AI model trained itself without their permission, be eligible for any royalty?

The Million-Dollar Question Is, Does Your AI-Generated Work Get Copyright Protection Right Now?

Should you find yourself engrossed in the creation of content through AI tools, it is imperative to understand a fundamental truth: you hold no ownership over it.

Yes, the captivating images meticulously crafted using Midjourney's capabilities remain bereft of protection. Startling as it may seem, anyone can readily seize hold of your creations, printing them on various articles such as t-shirts or book covers, and proceed to sell them without encountering any legal repercussions.

Likewise, for companies delving into the realm of AI-assisted content generation for written documents, articles, or news items, it is crucial to grasp the unsettling reality that such works possess susceptibility to being copied and utilized elsewhere.

Whether your AI endeavors revolve around fashion and interior design, advertising and marketing campaigns, the composition of music or sound effects, the crafting of 3D models, the creation of magazine covers or billboards, or even the development of educational content, it is vital to comprehend that copyright laws provide no protective haven for these endeavors.

Moreover, even the realm of AI-generated videos, such as the notable example of Netflix employing background animations in an anime series, remains beyond the realm of ownership, thereby relinquishing any claims to their authorship. By delving into the intricate facets of AI content ownership, we shed light on a matter of utmost importance, equipping you with the knowledge essential to navigate this intricate landscape.

Ongoing AI Innovations and Their IP Dilemmas

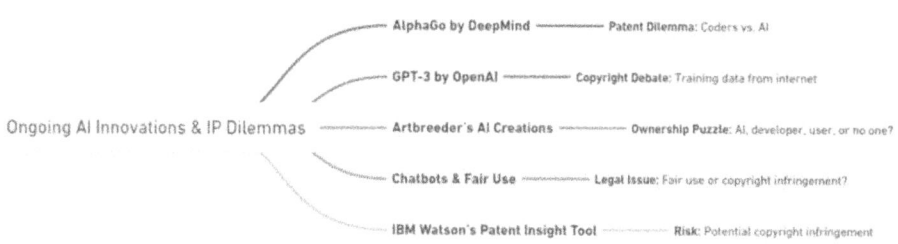

CHAPTER 5   AI COPYRIGHT AND INTELLECTUAL PROPERTY

AlphaGo by DeepMind

Google's DeepMind amazed the world with AlphaGo, an AI that conquered a human Go champion. The big question: who deserves the patent – the coders or the AI?

GPT-3 by OpenAI

OpenAI's GPT-3 can craft human-like text for various purposes. Its training on vast Internet data sparks debates about potential copyright issues.

Artbreeder's AI Creations

Artbreeder leverages AI to generate unique art pieces. The puzzle: who holds the copyright – the AI, its developer, the user, or no one?

Chatbots and Fair Use

Chatbots, using existing information, engage with customers. The issue: does this usage qualify as fair use or copyright infringement?

IBM Watson's Patent Insight Tool

IBM Watson aids companies in navigating the patent landscape but might inadvertently use copyrighted material, leading to potential infringement concerns.

> *But there is a catch; you can get copyright on AI generated work you produced if...*

CHAPTER 5   AI COPYRIGHT AND INTELLECTUAL PROPERTY

In the realm of creative rights, the United States has a strict policy: works birthed solely by machines don't qualify for copyright protection. Yet, a glimmer of hope shines when human involvement is substantial, opening the door to potential copyright claims. An intriguing case arose in September 2022 when the US Copyright Office granted registration to a comic book crafted with text-to-image AI. However, the verdict remains in suspense as this decision undergoes review. The extent of human input will hold immense sway in determining the fate of copyright eligibility.

Across the pond, in the vast expanse of the European Union, Giorgio Franceschelli, a computer scientist well-versed in AI copyright matters, emphasizes the significance of measuring human contribution, particularly in this domain. Meanwhile, the United Kingdom takes a distinct approach, standing alongside only a handful of nations that bestow copyright protection to works generated solely by computers. The intricacy lies in defining the author, as they are deemed "the person by whom the arrangements necessary for the creation of the work are undertaken," inviting diverse interpretations.

Also, it's worth noting that, obtaining copyright registration merely marks the initial stride, leaving the final judgment to the hands of the court for enforcement. As the fascinating world of generative AI continues to advance, legal experts must skillfully navigate these intricate complexities to safeguard the rights of creators and users alike. They bear the weight of responsibility, ensuring a harmonious symphony where creativity thrives, and protection abounds.

Defining the AI IP and Copyright Landscape

In the mesmerizing world of AI, where innovation dances hand in hand with uncertainty, a legal storm is brewing. As companies raking in profits from this transformative technology entrench themselves, the battle

CHAPTER 5   AI COPYRIGHT AND INTELLECTUAL PROPERTY

lines are drawn. On one side, they confidently proclaim their actions as lawful, while on the other, copyright holders stake their claims without committing to definitive action.

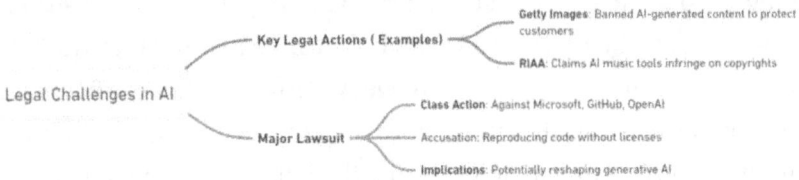

A tremor rippled through the creative sphere when Getty Images, recognizing the potential legal risks to its esteemed customers, took a drastic step: banning AI-generated content. Not long after, the music industry trade organization, RIAA, joined the fray, asserting that AI-powered music mixers and extractors infringed upon the cherished copyrights held by its members. The stage was set for a war of legal proportions.

But the opening act in the AI copyright wars arrived unexpectedly. A proposed class action lawsuit aimed at Microsoft, GitHub, and OpenAI unleashed shockwaves across the industry. The lawsuit accuses these giants of reproducing open-source code through their AI coding assistant, Copilot, without obtaining the necessary licenses. The implications of this case reverberate throughout the generative AI realm, holding the potential to shake its very foundation. Yet, the outcome remains uncertain – a cliffhanger that keeps us guessing.

Although legal actions have started to brew up, they are still astonishingly fewer than what was expected. Perhaps the weighty burden of litigation costs and the labyrinthine path to justice deter those most impacted by this technology – artists without the vast resources required to launch legal battles. In stark contrast, stock image sites stand on more solid ground, armed with proof that a substantial portion of their precious corpus was employed to train these AI models. With their financial might, they possess the means to drag their grievances to the courtroom. It seems that a few high-profile legal battles may be needed to unravel the tangled legal landscape that surrounds generative AI.

Proposed US and EU Copyright Laws and Regulations

In 2020, the United States House of Representatives passed the Algorithmic Accountability Act, which would require large technology companies to develop and implement algorithms that are fair, transparent, and accountable. The bill has not yet been passed by the Senate.

In 2021, the European Commission published a proposal for a new directive on copyright in the digital single market. The directive would address the challenges posed by AI-generated content, including the issue of who should own the IP rights to AI-generated content.

In summary, in the realm of AI-generated content, a landscape marked by novelty and ongoing exploration, the intricacies of copyright laws can indeed appear convoluted. Understandably so, for generative AI represents a burgeoning technology, still in the process of being comprehended and assimilated.

As generative AI expands into corporations, creative agencies, and personal projects, it's crucial to understand the legal framework governing it.

CHAPTER 5   AI COPYRIGHT AND INTELLECTUAL PROPERTY

AI Copyright and IP Consideration Checklist for Board and C-Suite

General Compliance

1. **Lawfulness and Transparency** (Score: 0-5 each)
   - Are our AI data processing activities compliant with relevant laws?
   - Is information about data usage communicated transparently to all stakeholders?

2. **Purpose and Data Limitation** (Score: 0-5 each)
   - Is data collection aligned with specific, legitimate purposes?
   - Are there measures in place to ensure data is not retained longer than necessary?

Authorship and Inventorship

3. **AI-Created Content** (Score: 0-5 each)
   - Are we prepared to address legal questions regarding AI as a potential inventor or author?
   - Do we have policies in place for allocating IP rights for AI-generated outputs?

Data Usage and Copyright

4. **Copyright and Data Training** (Score: 0-5 each)

   - Do we ensure the use of copyrighted data for AI training falls under fair use?

   - Are there mechanisms to prevent copyright infringement in our AI training processes?

Predictive Capabilities and IP Strategy

5. **AI in IP Strategy** (Score: 0-5 each)

   - Are we leveraging AI's predictive capabilities for forecasting IP trends?

   - Do we integrate AI insights into our overall IP strategy and decision-making processes?

6. **IP Management Tools** (Score: 0-5 each)

   - Are we using AI tools to efficiently track and manage IP rights and detect infringement?

   - Do we have automated systems to monitor and address potential IP violations?

Legal Framework and Compliance

7. **Legal Framework Understanding** (Score: 0-5 each)

   - Are we up-to-date with the evolving legal frameworks surrounding AI and IP?

   - Do we engage legal experts to navigate the complexities of AI-related IP issues?

CHAPTER 5 AI COPYRIGHT AND INTELLECTUAL PROPERTY

Data Security and Breach Response

8. **Data Security** (Score: 0-5 each)
   - Do we have robust measures in place to secure data used in AI processes?
   - Are there protocols for detecting, reporting, and mitigating data breaches?

Continuous Monitoring and Improvement

9. **Regular Audits** (Score: 0-5 each)
   - Do we conduct regular audits to ensure compliance with AI and IP regulations?
   - Is there a system for continuous monitoring and updating of our AI and IP policies?

10. **Training and Awareness** (Score: 0-5 each)
    - Do we provide regular training on AI and IP compliance to our employees?
    - Is there a culture of awareness about the importance of protecting IP rights?

Scoring and Interpretation

Each question is scored from 0 to 5, with a maximum possible score of 100 across all 20 questions.

- **0–40**: High risk – Immediate action required to address significant gaps in AI and IP risk management.

- **41–70**: Moderate risk – Areas for improvement identified, and a structured plan is needed to enhance AI and IP governance.

- **71–100**: Low risk – Robust AI and IP risk management practices in place, but continuous monitoring and minor adjustments are recommended.

Threshold for Passing

Organizations should aim for a minimum score of **70** to ensure they have adequately addressed key AI copyright and IP considerations and are prepared to manage related risks effectively.

Summary

As AI's creative prowess surges, we find ourselves in a legal quagmire – who truly owns the fruits of artificial intelligence? The current copyright landscape, designed for human creators, creaks under the weight of machine-generated content, leaving a vast expanse of creative works in a precarious legal limbo. From AlphaGo's strategic brilliance to GPT-3's linguistic gymnastics, each AI innovation births new intellectual property conundrums, challenging the very foundations of creativity and ownership. As legal battles brew and legislation struggles to keep pace, one thing is crystal clear: the intersection of AI and intellectual property law is not just a legal battlefield, but a crucible where the future of human-AI collaboration will be forged. In this brave new world of silicon-powered creativity, will we redefine authorship, or find ourselves mere spectators to an AI-dominated artistic revolution?

CHAPTER 6

AI Strategy

Creating an enterprise-wide AI strategy that aligns with and enhances the core business strategy is an iterative process and requires ongoing efforts. It's certainly not a one-time endeavor. Organizations must adopt a flexible operating model to assess and adapt their AI strategy in response to evolving market dynamics and technological advancements.

As the core business strategy and AI capabilities mature, leaders must continually refine their Objectives and Key Results (OKRs). More importantly, the leadership mindset should be to go beyond mere competitiveness and embrace AI and ML as key differentiators of their organization in the market. By staying ahead of the curve and leveraging AI effectively, organizations can achieve sustainable differentiation and gain a huge competitive edge.

Defining Your AI Strategy

To define your AI strategy, look at your business strategy. And then work backwards from there.

Leaders often overlook the fact that developing AI solutions requires more than just prioritizing isolated use cases, an approach which most organizations seem to be taking. The true transformation potential lies in a connected and coordinated AI strategy that seamlessly integrates with the overarching business strategy. Unfortunately, many organizations make the mistake of prioritizing use cases and then leaving the AI strategy solely in the hands of IT or data science leadership, which can hinder their ability to leverage AI for customer engagement, product innovation, and operational excellence.

The most successful AI strategies start by focusing on the core business strategy, rather than explicitly mentioning AI. Through collaboration with engaged cross-functional leaders across various business divisions and involving workers at all levels, AI strategy becomes the driving force that aligns with key performance indicators (KPIs) to foster competitive advantage and growth.

Subsequently, these localized plans need to be integrated with the core business strategy and metrics. This crucial step ensures that mutual goals and initiatives are aligned, allowing AI to permeate throughout the organization and generate the desired efficiency and value-driven outcomes that sustain ongoing returns.

Avoid Overemphasis on Cost Savings

Organizations aiming for success need a combination of efficiency and value creation. While efficiency optimizes existing processes, AI also unlocks new business opportunities and capabilities in adjacent or untapped areas.

It's proven over and over again that lower-achieving organizations prioritize efficiency and cost reduction, while high-achievers focus on growth-oriented goals like improving customer satisfaction, introducing new products, and expanding into new markets. High achievers maintain a growth mindset and seize opportunities that those solely focused on efficiency might overlook.

CHAPTER 6   AI STRATEGY

Communicating a Clear AI Strategy Elevates Market Value

In high-achieving organizations, top leadership plays a crucial role as the primary communicator of AI initiatives. It's been seen time and again that organizations with a clear vision communicated by their leaders are orders of magnitude more likely to achieve desired outcomes compared to those lacking such clarity. And the reverse is also true as we all saw in the case of Silicon Valley Bank.

Effective leaders go beyond merely communicating clearly and advocacy; they also emphasize the implications and trade-offs their teams will have to make to embark on the AI journey. Maintaining laser focus and alignment throughout the organization requires leaders to clarify the vision and its implications at all levels and repeatedly.

On communicating AI vision publicly, leaders should recognize the influence they have in shaping market perceptions and attracting investors. Publicly communicating the company's vision will amplify excitement around the organization, signaling to the capital and talent markets that the organization is dedicated to an ambitious and promising future.

On the contrary, if the vision does not warrant such a resounding call for change, there is a risk that the gravitational pull toward the status quo may hinder even the most robust strategy's outcomes.

Mindset for a Robust AI Strategy

In summary, for organizations looking to make investment in AI will have to commit for the long haul since they most likely will have to compete fiercely for AI talent, change their operating models and de-prioritize some initiatives to get focus on AI. It's also crucial to understand.

CHAPTER 6   AI STRATEGY

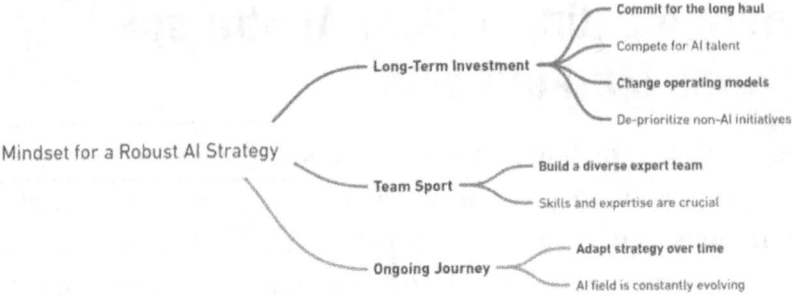

- **AI is a long-term investment.** It takes time to develop and deploy AI systems. Organizations should be prepared to invest in AI for the long term.

- **It's a team sport.** The development and deployment of AI systems requires a team of experts with different skills and expertise. Organizations should build a team that has the skills and expertise needed to succeed in AI.

- **It's a journey, not a destination.** The field of AI is constantly evolving. Organizations should be prepared to adapt their strategy and roadmap as AI technology continues to develop.

Establishing a versatile and robust artificial intelligence (AI) strategy is essential for businesses aiming to maintain competitive advantage and operational excellence. This comprehensive chapter delves deeply into the nuances of forming an AI strategy that spans legal readiness, strategic deployment, and ethical considerations, while focusing on generating tangible business value.

CHAPTER 6 AI STRATEGY

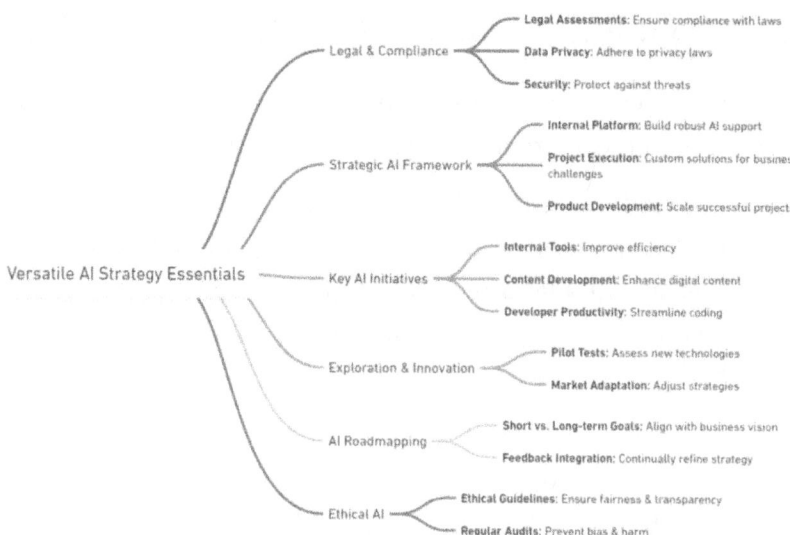

1. Setting the Legal and Compliance Groundwork

Before diving into AI integration, a business must rigorously address the legal, data privacy, and security concerns associated with new technologies:

- **Legal Assessments:** Thorough legal evaluations are essential for new models and third-party solutions to ensure they comply with existing laws and regulations.

- **Data Privacy Assessments:** This involves detailed scrutiny of how data is handled, ensuring adherence to privacy laws and reducing the risk of data breaches.

- **Security Assessments:** Assessing the security measures of new AI tools and models to protect against vulnerabilities and threats is crucial.

2. Strategic Framework for AI Implementation

A strategic framework lays out the path for AI integration, highlighting the need for a structured approach to operationalize AI within the enterprise:

- **Internal AI Platform Development:** Building a robust internal platform that supports AI deployment is critical. This platform should support scalable, secure architecture and integrate seamlessly with existing business processes.

- **AI Project Execution:** Custom AI projects should be developed to address unique business challenges, leveraging the capabilities of the internal AI platform to enhance efficiency and effectiveness.

- **AI Product Development:** Transitioning successful AI projects into scalable products that can be marketed or integrated into broader business practices to maximize ROI.

3. Prioritizing Key AI Initiatives

Focusing on specific, high-impact AI initiatives can drive significant business value:

- **Internal Tools:** Develop AI-driven internal tools aimed at improving operational efficiency and decision-making.

- **Content Development:** Utilize AI for enhancing the creation and management of digital content, potentially increasing the speed and quality of content delivery.

- **Enhancing Developer Productivity:** Implement AI solutions such as coding assistants to streamline software development processes and improve output quality.

4. Exploration and Innovation in AI

Continuous exploration of emerging AI technologies and methodologies is vital to stay ahead in a fast-paced market:

- **Pilot and Testing:** Running pilot tests for new AI technologies to assess their impact and scalability before full deployment.
- **Market Adaptation:** Adapting AI strategies based on market feedback and technological advancements to ensure continuous improvement and relevance.

5. AI Strategy Roadmapping

A well-defined AI roadmap is essential for guiding the phased deployment of AI projects and ensuring alignment with long-term business goals:

- **Short-term vs. Long-term Goals:** Delineate between immediate AI goals and long-term aspirations, ensuring resources are appropriately allocated.
- **Feedback Integration:** Incorporate feedback mechanisms to continually refine and adjust the AI strategy as needed.

CHAPTER 6   AI STRATEGY

6. Ethical AI Deployment

Ensuring the responsible use of AI is a cornerstone of any strategy:

- **Ethical Guidelines:** Develop and enforce guidelines for ethical AI use that emphasize fairness, transparency, and accountability.

- **Regular Audits:** Conduct regular audits of AI applications to ensure they do not inadvertently propagate biases or cause harm.

Navigating the Buy vs. Build Decision in AI Strategy

In the realm of AI integration within enterprises, one of the most pivotal decisions organizations must make is whether to build their own AI solutions or buy existing products. This chapter explores the nuanced decision-making process involved in the "buy vs. build" dilemma, using a tiered approach to AI adoption as a framework for understanding when each option might be most beneficial.

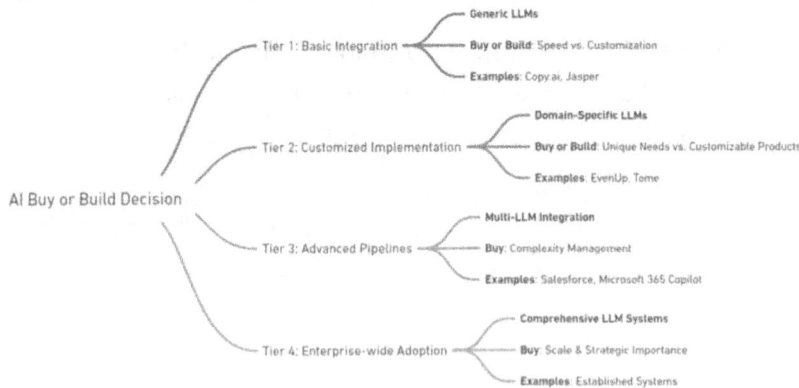

CHAPTER 6   AI STRATEGY

The complexity and scope of AI needs within an organization can be categorized into different tiers, each representing a level of AI sophistication and application:

> **Tier 1: Basic LLM Integration**
>
> **Tier 2: Customized LLM Implementation**
>
> **Tier 3: Advanced LLM Pipelines**
>
> **Tier 4: Enterprise-wide LLM Adoption**

Each tier requires a different approach to the buy vs. build question, influenced by factors such as cost, expertise, strategic importance, and implementation time.

Tier 1: Basic LLM Integration

- **Characteristics:** Utilization of generic large language models (LLMs) via API calls for enhancing productivity tools.

Decision Factors:

- **Build and Buy:** Smaller enterprises or startups may choose to build their basic integrations to tailor specific functions or buy to speed up deployment.

- **Example Companies:** Companies like Copy.ai and Jasper illustrate the effectiveness of integrating basic LLM functionalities.

Tier 2: Customized LLM Implementation

- **Characteristics**: Fine-tuned and domain-specific LLMs that address particular departmental challenges or enhance domain-specific workflows.

Decision Factors:

- **Build and Buy**: Organizations with unique needs that generic products cannot meet might opt to build. However, buying can also be considered if there are customizable products available that can be adapted to fit specific needs.

- **Example Companies**: EvenUp and Tome are examples where custom solutions provide significant advantages over off-the-shelf products.

Tier 3: Advanced LLM Pipelines

- **Characteristics**: Integration of multiple LLMs for complex, multistep use cases spanning several departments.

Decision Factors:

- **Buy**: The complexity of managing multiple integrated LLMs generally makes buying more feasible, as it requires extensive expertise and continuous upkeep which can be resource-intensive.

- **Example Companies**: Salesforce and Microsoft 365 Copilot demonstrate the efficiency of buying sophisticated, integrated AI solutions that are too complex to build in-house effectively.

Tier 4: Enterprise-wide LLM Adoption

- **Characteristics**: Comprehensive LLM systems that provide extensive analytics, reasoning capabilities, and out-of-the-box connectors for large-scale, strategic enterprise applications.

Decision Factors:

- **Buy**: The scale and strategic importance of enterprise-wide solutions typically make the purchase of established systems more practical. Building such extensive systems in-house would require significant investment and risk.

AI Strategy Consideration Checklist for Board and C-Suite

Creating a robust AI strategy is critical for maintaining competitive advantage and operational excellence. This checklist provides key questions for board members and C-suite executives to consider, ensuring a comprehensive approach to AI strategy. Each question is scored from 0 to 5, with a total possible score of 90.

Scoring System

- **0 points:** Not addressed
- **1 point:** Partially addressed
- **2 points:** Fully addressed, but needs improvement
- **3 points:** Fully addressed and well-executed

CHAPTER 6   AI STRATEGY

Setting the Legal and Compliance Groundwork

1. **Legal Assessments**
   - Have we conducted thorough legal evaluations for new AI models and third-party solutions to ensure compliance with existing laws and regulations?
     - Score: 0-3
2. **Data Privacy Assessments**
   - Do we scrutinize data handling processes to ensure adherence to privacy laws and reduce the risk of data breaches?
     - Score: 0-3
3. **Security Assessments**
   - Have we assessed the security measures of new AI tools and models to protect against vulnerabilities and threats?
     - Score: 0-3

Strategic Framework for AI Implementation

4. **Internal AI Platform Development**
   - Have we built a robust internal platform that supports scalable and secure AI deployment, integrating seamlessly with existing business processes?
     - Score: 0-3

5. **AI Project Execution**

    - Are we developing custom AI projects to address unique business challenges, leveraging our internal AI platform to enhance efficiency and effectiveness?

        - Score: 0-3

6. **AI Product Development**

    - Do we transition successful AI projects into scalable products that can be marketed or integrated into broader business practices to maximize ROI?

        - Score: 0-3

Prioritizing Key AI Initiatives

7. **Internal Tools Development**

    - Are we developing AI-driven internal tools aimed at improving operational efficiency and decision-making?

        - Score: 0-3

8. **Content Development**

    - Are we utilizing AI to enhance the creation and management of digital content, potentially increasing the speed and quality of content delivery?

        - Score: 0-3

9. **Enhancing Developer Productivity**
    - Have we implemented AI solutions such as coding assistants to streamline software development processes and improve output quality?
    - **Score: 0-3**

Exploration and Innovation in AI

10. **Pilot and Testing**
    - Are we running pilot tests for new AI technologies to assess their impact and scalability before full deployment?
    - **Score: 0-3**

11. **Market Adaptation**
    - Do we adapt our AI strategies based on market feedback and technological advancements to ensure continuous improvement and relevance?
    - **Score: 0-3**

AI Strategy Roadmapping

12. **Short-Term vs. Long-Term Goals**
    - Have we delineated between immediate AI goals and long-term aspirations, ensuring resources are appropriately allocated?
    - **Score: 0-3**

13. **Feedback Integration**
    - Do we incorporate feedback mechanisms to continually refine and adjust our AI strategy as needed?
      - Score: 0-3

Ethical AI Deployment

14. **Ethical Guidelines**
    - Have we developed and enforced guidelines for ethical AI use that emphasize fairness, transparency, and accountability?
      - Score: 0-3

15. **Regular Audits**
    - Do we conduct regular audits of AI applications to ensure they do not inadvertently propagate biases or cause harm?
      - Score: 0-3

Leadership and Communication

16. **Leadership Advocacy**
    - Are top leaders actively communicating and advocating the AI vision both internally and externally?
      - Score: 0-3

CHAPTER 6 AI STRATEGY

17. **Vision Communication**

    - Do leaders emphasize the implications and trade-offs required for the AI journey and clarify the vision at all levels?

    - Score: 0-3

18. **Public Communication**

    - Are we effectively communicating our AI vision publicly to shape market perceptions and attract investors?

    - Score: 0-3

Commitment and Long-Term Investment

19. **Long-term Investment**

    - Are we prepared to invest in AI for the long term, understanding it takes time to develop and deploy AI systems?

    - Score: 0-3

20. **Team Building**

    - Have we built a team of experts with the necessary skills and expertise to succeed in AI development and deployment?

    - Score: 0-3

Scoring and Interpretation

- **0–40**: High risk – Immediate action required to address significant gaps in AI strategy.

- **41–70**: Moderate risk – Areas for improvement identified, and a structured plan is needed to enhance AI strategy.

- **71–90**: Low risk – Robust AI strategy practices in place, but continuous monitoring and minor adjustments are recommended.

Threshold for Passing

Organizations should aim for a minimum score of **70** to ensure they have adequately addressed key AI strategy considerations and are prepared to manage AI initiatives effectively.

Summary

Developing a comprehensive AI strategy involves balancing rigorous legal and security assessments, strategic implementation frameworks, and ethical usage guidelines. By focusing on specific, value-driven initiatives and maintaining flexibility to adapt to new challenges and opportunities, enterprises can effectively leverage AI to drive innovation and operational efficiency. This proactive and structured approach ensures that AI initiatives are both impactful and sustainable, aligning with broader business objectives and fostering an environment of continuous improvement and ethical responsibility.

CHAPTER 7

AI Operating Model

Creating an effective operating model for data and AI integration is not only crucial but one of the first steps needed for any material results, let alone sustained business outcomes.

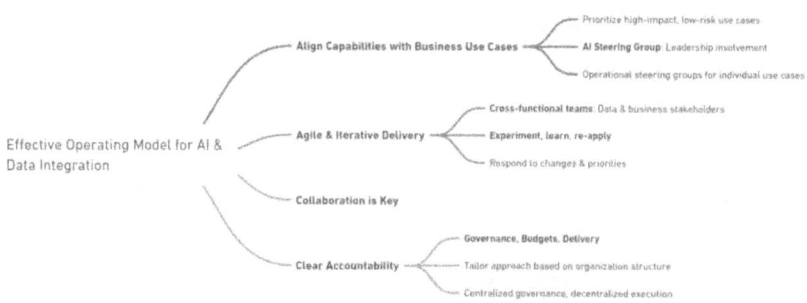

In the context of AI and data, it starts with aligning specific data and AI capabilities with prioritized business use cases identified in the strategic assessment. An organization might have some of these capabilities already but need others. To ensure the focus on essential high-impact, low-risk use cases, business leaders should establish an AI steering group or include data and AI development in existing leadership team meetings, with the top leadership leading the agenda. Additionally, individual use-case areas should have their own operational steering groups.

Another non-negotiable element of operating model is to utilize an agile and iterative delivery approach with cross-functional participation (i.e., both data and business stakeholders) which can respond to changes and business priorities while still letting data product teams to experiment, learn, and re-apply learnings in iterations.

CHAPTER 7   AI OPERATING MODEL

Close collaboration between data and business functions should be a conversation starter to achieving any tangible and sustainable results. For instance, in marketing, achieving personalized and data-driven campaigns requires collaboration across various areas, including data availability, content production, customer treatment models, and channel strategies. Similar collaboration is imperative for process automation, where predictive models require aligned technical systems and timely intervention.

Reference: MIT Press

The third and a critical piece in having at least a foundational operating model in place is to assign clear, accountability for governance, budgets and delivery across your teams.

There are numerous approaches organizations can take, all of which depends on their existing organization structures and their track record on innovation.

The AI Operating Model Framework

The AI operating model framework illustrated provides a structured approach to organizing AI-related activities across the enterprise. This model ensures that AI initiatives are well-coordinated, secure, and aligned with strategic business objectives.

CHAPTER 7   AI OPERATING MODEL

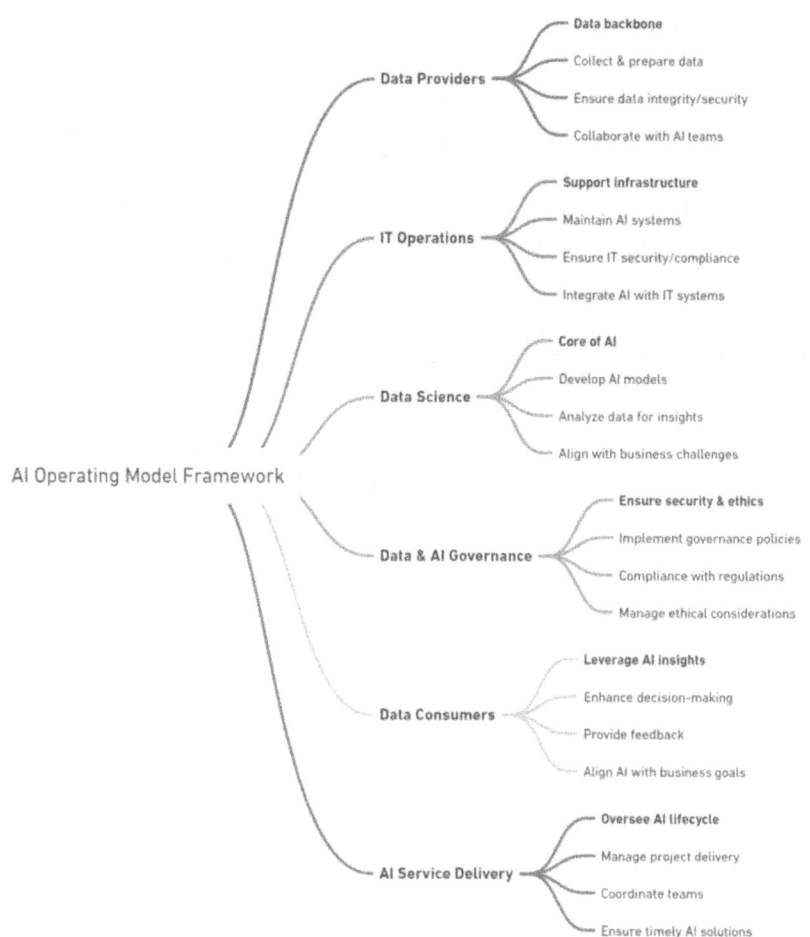

1. Data Providers

Data providers form the backbone of the AI operating model, supplying the necessary data for AI and analytics. These include various departments such as finance, sales, HR, production, and IT. Local business and IT teams within these departments play a critical role in ensuring data availability and quality.

Key Responsibilities

- Collecting and preparing data for AI applications
- Ensuring data integrity and security
- Collaborating with centralized data and AI teams to meet data needs

2. IT Operations

IT operations are responsible for maintaining the technological infrastructure that supports AI initiatives. This includes managing data storage, processing capabilities, and IT security measures.

Key Responsibilities

- Providing and maintaining the infrastructure for AI systems
- Ensuring robust IT security and compliance
- Facilitating seamless integration of AI solutions with existing IT systems

3. Data Science

The data science team is at the core of the AI operating model, developing and deploying AI models and algorithms. This centralized team works closely with both IT operations and business units to create effective AI solutions.

Key Responsibilities

- Developing and fine-tuning AI models and algorithms
- Conducting data analysis to derive actionable insights
- Collaborating with business units to ensure AI solutions address real-world challenges

4. Data and AI Governance

Strong governance is essential for ensuring that AI initiatives are secure, ethical, and compliant with regulations. The governance team oversees data management, ethical guidelines, and compliance frameworks.

Key Responsibilities

- Implementing data governance policies and procedures
- Ensuring compliance with data protection regulations
- Managing ethical considerations in AI deployment

5. Data Consumers

Data consumers are the end users of AI-generated insights, including process owners, managers, business analysts, and C-suite executives. These users leverage AI insights to make informed decisions and drive business strategies.

Key Responsibilities

- Utilizing AI insights to enhance decision-making
- Providing feedback to data science and IT teams for continuous improvement
- Ensuring that AI solutions align with business objectives and needs

6. AI Service Delivery Management

AI service delivery management oversees the end-to-end delivery of AI services, ensuring that AI initiatives are executed effectively and deliver the expected business value.

CHAPTER 7   AI OPERATING MODEL

Key Responsibilities

- Managing the lifecycle of AI projects from inception to deployment
- Coordinating between data providers, IT operations, data science, and data consumers
- Ensuring timely and efficient delivery of AI solutions

Implementing the AI Operating Model

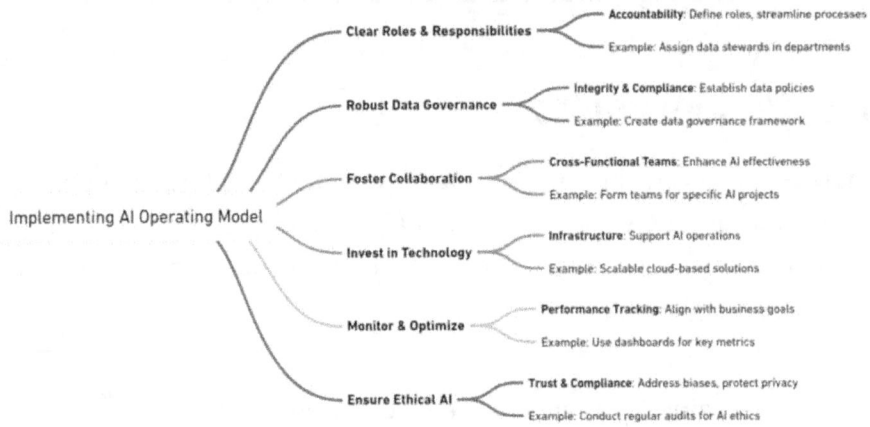

To implement a successful AI operating model, organizations should follow these strategic steps:

1. Establish Clear Roles and Responsibilities

Defining clear roles and responsibilities for each component of the AI operating model is essential. This clarity ensures accountability and streamlines processes across the organization.

Example: Assign specific data stewards within each department to oversee data quality and collaboration with centralized teams.

2. Develop Robust Data Governance Policies

Implementing robust data governance policies is critical for maintaining data integrity, security, and compliance. This includes establishing data management protocols, ethical guidelines, and regulatory compliance measures.

Example: Develop a data governance framework that outlines data access controls, usage policies, and compliance requirements.

3. Foster Collaboration Across Teams

Encouraging collaboration between data providers, IT operations, data science, and data consumers is crucial for the success of AI initiatives. Cross-functional teams can provide diverse perspectives and expertise, enhancing the effectiveness of AI solutions.

Example: Create cross-functional project teams to work on specific AI initiatives, ensuring representation from all relevant departments.

4. Invest in Technology and Infrastructure

Investing in the right technology and infrastructure is essential for supporting AI operations. This includes data storage solutions, processing capabilities, and advanced analytics tools.

Example: Implement scalable cloud-based infrastructure to support the growing data and computational needs of AI projects.

5. Monitor and Optimize Performance

Continuous monitoring and optimization of AI systems ensure that they remain effective and aligned with business goals. This involves tracking performance metrics, identifying areas for improvement, and making necessary adjustments.

Example: Use performance dashboards to monitor key metrics such as model accuracy, data processing times, and user satisfaction.

6. Ensure Ethical AI Practices

Maintaining ethical AI practices is critical for building trust and ensuring compliance. This includes addressing biases, ensuring transparency, and protecting user privacy.

Example: Conduct regular audits to identify and mitigate biases in AI models, and establish clear guidelines for ethical AI use.

Strategic Implications for Executives and Boards

Implementing an effective AI operating model has several strategic implications for corporate boards and senior executives:

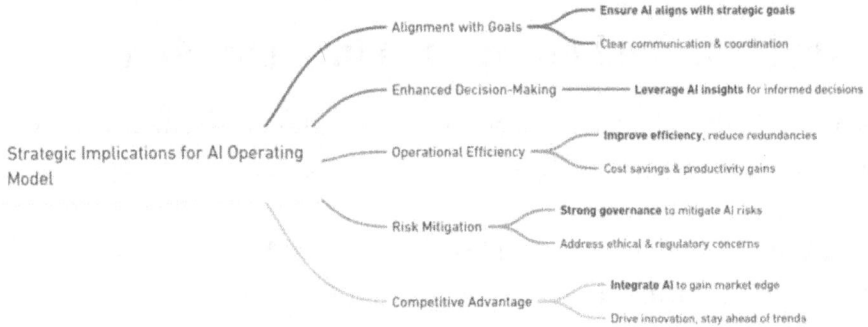

1. Alignment with Strategic Goals

Ensuring that AI initiatives are aligned with the organization's strategic goals is essential for driving business value. This requires clear communication and coordination across all levels of the organization.

2. Enhanced Decision-Making

Leveraging AI-generated insights enhances decision-making capabilities, providing executives with the information needed to make informed strategic choices.

3. Operational Efficiency

A well-structured AI operating model improves operational efficiency, streamlining processes and reducing redundancies. This leads to cost savings and improved productivity.

4. Risk Mitigation

Implementing strong governance and compliance measures helps mitigate risks associated with AI deployment, including data breaches, ethical concerns, and regulatory penalties.

5. Competitive Advantage

By effectively integrating AI into business operations, organizations can gain a competitive advantage, driving innovation and staying ahead of industry trends.

Who Owns Data and AI Budgets?

In the initial years of establishing the CoE, centralizing budgets proves advantageous. This centralization enables prioritization and scalability of data and AI activities. Often, individual business units hesitate to bear the costs of company-wide capability building, resulting in disconnected AI solutions.

CHAPTER 7   AI OPERATING MODEL

Lack of a common roadmap, clear governance, and prioritization can lead to resource allocation challenges, which is where a CoE approach can be extremely beneficial.

Another crucial practice for increasing the chances of success in AI initiatives is to ensure that all stakeholders receive equal incentives.

By setting shared goals for marketers, data scientists, and data engineers, such as increasing marketing campaign lift through AI-driven targeting, organizations gain crucial alignment to boost collective effort and maximize results.

Common Challenges in AI Operating Model Delivery at Scale

While tooling, integration and data issues are common, success is primarily influenced by enterprise-level considerations. Key aspects include:

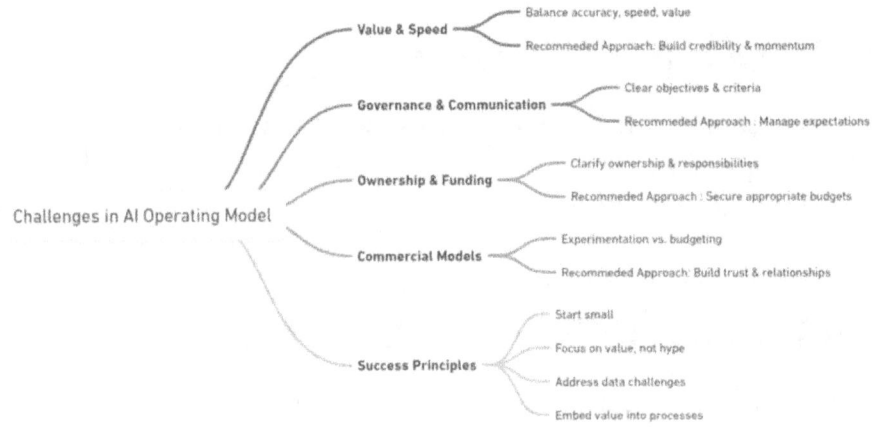

- **Value and delivery speed:** Balancing accuracy, speed, and value is crucial when translating business problems into AI use cases. Progress must be delivered in a way that builds credibility and generates momentum.

- **Governance and communications:** Data science involves exploring various options and dealing with uncertainties.

  Effective governance, clear objectives, success criteria, and regular communication are essential to manage stakeholder expectations.

- **Ownership, funding, and commercials:** Successful AI projects require multidisciplinary teams that bridge IT and the business. Clarity of ownership and funding responsibilities is crucial, along with determining who ultimately owns the benefits.

- **Commercial models and budgeting:** AI and data science projects entail experimentation, but commercial models and budgets are often assessed on a case-by-case basis. Building good relationships and trust can help secure an appropriate budget for the delivery team.

Additional principles that influence project success include starting small, focusing on value rather than hype, addressing data challenges, embedding value into business processes, and following productionization processes similar to traditional software application lifecycles.

CHAPTER 7   AI OPERATING MODEL

AI Operating Model Checklist for Leaders

To ensure an effective AI operating model, board members and C-suite executives should ask the following questions. This checklist incorporates a scoring system to quantitatively assess the robustness of the AI operating model.

Scoring System

- **0 points:** Not addressed
- **1 point:** Partially addressed
- **2 points:** Fully addressed, but needs improvement
- **3 points:** Fully addressed and well-executed

Alignment with Business Use Cases

1. **Strategic Alignment** (Score: 0-3 each)
   - Are AI and data capabilities aligned with prioritized business use cases identified in the strategic assessment?
   - Is there an AI steering group or are AI development discussions included in leadership team meetings?

Agile and Iterative Approach

2. **Agile Methodology** (Score: 0-3 each)
   - Are agile and iterative delivery approaches utilized with cross-functional participation?
   - Can data product teams experiment, learn, and re-apply learnings in iterations?

Collaboration and Integration

3. **Cross-functional Collaboration** (Score: 0-3 each)
   - Is there close collaboration between data and business functions to achieve tangible results?
   - Are specific areas like marketing and process automation benefiting from this collaboration?

Accountability and Governance

4. **Clear Accountability** (Score: 0-3 each)
   - Are there clear accountabilities for governance, budgets, and delivery across teams?
   - Is there a structured approach for assigning roles and responsibilities within the AI operating model?

CHAPTER 7   AI OPERATING MODEL

Data and IT Infrastructure

5. **Data Availability and Quality** (Score: 0-3 each)
   - Are data providers ensuring the availability and quality of data for AI applications?
   - Are IT operations providing and maintaining robust infrastructure for AI systems?

6. **IT Security and Compliance** (Score: 0-3 each)
   - Are IT security and compliance measures robust and effective?
   - Is there seamless integration of AI solutions with existing IT systems?

Data Science and AI Development

7. **Model Development and Deployment** (Score: 0-3 each)
   - Is the data science team effectively developing and fine-tuning AI models and algorithms?
   - Are AI models addressing real-world business challenges and deriving actionable insights?

Governance and Ethical Considerations

8. **AI Governance Policies** (Score: 0-3 each)
   - Are data governance policies and procedures well-implemented?

- Is there strong oversight to ensure compliance with data protection regulations and manage ethical considerations in AI deployment?

Service Delivery and Performance Monitoring

9. **AI Service Delivery Management** (Score: 0-3 each)
   - Is there efficient management of the AI project lifecycle from inception to deployment?
   - Are AI services delivered effectively and aligned with business expectations?

10. **Performance Monitoring and Optimization** (Score: 0-3 each)
    - Are AI systems continuously monitored and optimized for performance and alignment with business goals?
    - Are performance metrics tracked, and areas for improvement identified and addressed?

Ethical AI Practices

11. **Ethical AI Practices** (Score: 0-3 each)
    - Are there guidelines and regular audits to identify and mitigate biases in AI models?
    - Are AI practices transparent, ensuring privacy and ethical considerations?

CHAPTER 7   AI OPERATING MODEL

Strategic Implementation

12. **Clear Roles and Responsibilities** (Score: 0-3 each)

    - Are roles and responsibilities clearly defined for each component of the AI operating model?
    - Does this clarity ensure accountability and streamline processes?

13. **Collaboration Across Teams** (Score: 0-3 each)

    - Is cross-functional collaboration encouraged to integrate AI initiatives seamlessly into business operations?
    - Are project teams representing all relevant departments?

14. **Technology and Infrastructure Investment** (Score: 0-3 each)

    - Is there investment in the right technology and infrastructure to support AI operations?
    - Are scalable cloud-based infrastructure and advanced analytics tools implemented?

15. **Continuous Monitoring and Optimization** (Score: 0-3 each)

    - Are AI systems continuously monitored for performance optimization?
    - Are there dashboards to track key metrics like model accuracy and processing times?

16. **Ethical AI Practices** (Score: 0-3 each)

    - Are there regular audits to identify and mitigate biases in AI models?

    - Are clear guidelines established for ethical AI use?

Total Scoring

Each question is scored from 0 to 3, with a maximum possible score of 48 across all 16 questions.

Interpretation

- **0-16**: High risk – Immediate action required to address significant gaps in the AI operating model.

- **17-32**: Moderate risk – Areas for improvement identified, and a structured plan is needed to enhance the AI operating model.

- **33-48**: Low risk – Robust AI operating model practices in place, but continuous monitoring and minor adjustments are recommended.

Threshold for Passing

Organizations should aim for a minimum score of **33** to ensure they have adequately addressed key considerations for an effective AI operating model and are prepared to manage related risks effectively.

CHAPTER 7   AI OPERATING MODEL

Summary

Operating Model is a foundational first step in expediting your organization's AI journey. Without a well-thought-out and communicated one, most organizations are simply signing a Loss/write off on their P/L right from the beginning.

CHAPTER 8

Determining AI Maturity for Your Organization

AI maturity refers to the level of sophistication in AI deployment and the corresponding increase in organizational capabilities and ROI. The maturity model, as illustrated in the provided image, outlines seven levels of AI sophistication, from foundational to advanced multiagent systems.

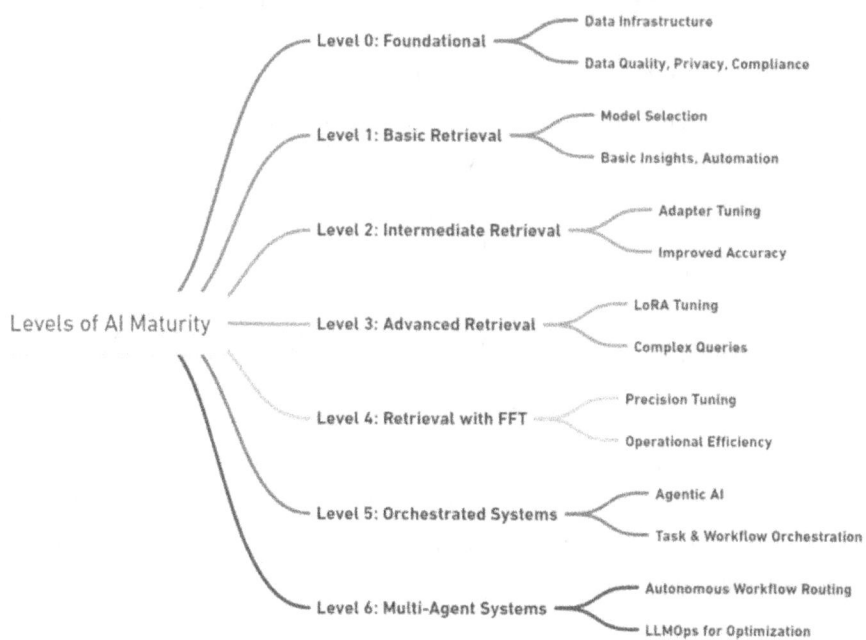

Levels of AI Maturity

Level 0: Foundational

At the foundational level, organizations begin by procuring, generating, curating, and preparing data. This stage focuses on building the necessary data infrastructure and ensuring data quality, privacy, and compliance. Establishing robust data management practices is critical for the success of subsequent AI initiatives.

Level 1: Basic Retrieval Augmentation

Organizations at this level start selecting models and prompts. Basic retrieval augmentation techniques are applied to enhance data retrieval processes. The primary objective is to serve models that can provide basic insights and automate simple tasks, laying the groundwork for more complex AI applications.

Level 2: Intermediate Retrieval Augmentation

In this stage, organizations continue to select models and prompts while advancing to intermediate retrieval augmentation. Techniques such as adapter tuning are employed to refine model outputs based on organizational data. This level focuses on improving the accuracy and relevance of AI-generated insights.

Level 3: Advanced Retrieval Augmentation

Advanced retrieval augmentation marks a significant leap in AI sophistication. Organizations tune their models using techniques like LoRA (Low-Rank Adaptation) tuning. These methods enable AI systems to handle more complex queries and provide deeper insights, enhancing decision-making processes.

Level 4: Advanced Retrieval Augmentation with FFT

This level builds on the previous stages by incorporating advanced tuning methods such as Fast Fourier Transform (FFT). These techniques allow for more precise adjustments to AI models, improving their performance and adaptability. The ability to serve highly refined models significantly boosts operational efficiency and effectiveness.

Level 5: Orchestrated Agentic Systems

At this level, AI systems become agentic, orchestrating tasks and workflows through large language models (LLMs). This stage involves grounding and evaluating AI outputs to ensure accuracy and reliability. Organizations can deploy AI systems to manage complex processes, reducing manual intervention and enhancing productivity.

Level 6: Multiagent Systems and Workflow Orchestration

The pinnacle of AI maturity involves the deployment of multiagent systems capable of orchestrating and routing workflows autonomously. These systems leverage advanced retrieval augmentation and sophisticated tuning techniques. Key components at this level include evaluation and observability frameworks and LLMOps (Large Language Model Operations) to maintain and optimize AI performance.

Responsible AI

Throughout the AI maturity journey, maintaining responsible AI practices is paramount. This includes ensuring security, privacy, compliance, and ethical considerations. As organizations progress through the maturity levels, they must continuously evaluate and refine their AI governance frameworks to address these critical issues.

Strategic Implications for Executives and Boards

Achieving AI maturity requires strategic oversight and a clear roadmap. Corporate boards and senior executives play a crucial role in guiding their organizations through the AI maturity model. Here are key strategic considerations:

1. Invest in Data Infrastructure

Building a robust data infrastructure is foundational for AI maturity. Investments in data management, privacy, and compliance are critical to support advanced AI applications.

2. Develop Incremental AI Capabilities

Organizations should adopt a phased approach to AI deployment, gradually advancing through the maturity levels. This ensures that each stage builds on the previous one, enhancing overall capabilities and ROI.

3. Foster Cross-Functional Collaboration

AI initiatives require collaboration across various business functions. Encouraging cross-functional teams to work together can accelerate AI adoption and ensure that AI solutions address comprehensive business challenges.

4. Prioritize Responsible AI Practices

Maintaining responsible AI practices is essential at every stage of AI maturity. Organizations must implement robust governance frameworks to ensure ethical AI deployment, data privacy, and regulatory compliance.

5. Continuously Evaluate and Optimize

As AI systems become more sophisticated, continuous evaluation and optimization are necessary. Implementing evaluation and observability frameworks ensures that AI systems remain effective and aligned with business goals.

Checklist for Determining AI Maturity for Your Organization

Scoring System

- **0 points:** Not addressed
- **1 point:** Partially addressed
- **2 points:** Fully addressed, but needs improvement
- **3 points:** Fully addressed and well-executed

Foundational Level (Level 0)

1. **Data Infrastructure Development** (Score: 0-3)
   - Have we established a robust data infrastructure to support AI initiatives?
   - Are data quality, privacy, and compliance measures effectively implemented?

CHAPTER 8   DETERMINING AI MATURITY FOR YOUR ORGANIZATION

2. **Data Management Practices** (Score: 0-3)
   - Do we have strong data management practices that ensure the reliability and accuracy of our data?

Basic Retrieval Augmentation (Level 1)

3. **Model and Prompt Selection** (Score: 0-3)
   - Have we started selecting appropriate models and prompts for basic AI tasks?
   - Are these models providing basic insights that automate simple tasks?

Intermediate Retrieval Augmentation (Level 2)

4. **Adapter Tuning** (Score: 0-3)
   - Are we employing adapter tuning techniques to refine AI model outputs?
   - Is the accuracy and relevance of AI-generated insights improving as a result?

Advanced Retrieval Augmentation (Level 3)

5. **LoRA Tuning** (Score: 0-3)
   - Have we implemented Low-Rank Adaptation (LoRA) tuning to handle more complex queries?
   - Are these techniques enhancing our decision-making processes?

CHAPTER 8   DETERMINING AI MATURITY FOR YOUR ORGANIZATION

Advanced Retrieval Augmentation with FFT (Level 4)

6. **Fast Fourier Transform (FFT) Techniques** (Score: 0-3)

    - Are we incorporating FFT techniques to make precise adjustments to our AI models?

    - Is there a noticeable improvement in AI performance and adaptability?

Orchestrated Agentic Systems (Level 5)

7. **AI System Grounding and Evaluation** (Score: 0-3)

    - Have we grounded and evaluated AI outputs to ensure accuracy and reliability?

    - Are our AI systems effectively managing complex processes with minimal manual intervention?

Multiagent Systems and Workflow Orchestration (Level 6)

8. **Deployment of Multiagent Systems** (Score: 0-3)

    - Are we deploying multiagent systems capable of autonomously orchestrating workflows?

    - Have we implemented evaluation and observability frameworks to optimize AI performance?

9. **LLMOps Implementation** (Score: 0-3)

    - Are Large Language Model Operations (LLMOps) in place to maintain and optimize AI systems?

- Do these operations ensure the sustainability and scalability of our AI initiatives?

Responsible AI Practices

10. **Ethical AI Deployment** (Score: 0-3)
    - Do we have a governance framework that ensures ethical AI deployment throughout the maturity journey?
    - Are we continuously evaluating and refining our AI governance to address security, privacy, and compliance issues?

Strategic Implications for Executives and Boards

11. **Investment in Data Infrastructure** (Score: 0-3)
    - Are we making the necessary investments in data infrastructure to support our AI maturity goals?

12. **Incremental AI Capability Development** (Score: 0-3)
    - Are we adopting a phased approach to AI deployment, advancing through maturity levels in a structured manner?

13. **Cross-Functional Collaboration** (Score: 0-3)
    - Are we fostering collaboration across business functions to accelerate AI adoption and address comprehensive challenges?

14. **Continuous Evaluation and Optimization**
    (Score: 0-3)

    - Do we have processes in place for continuous evaluation and optimization of AI systems?

Total Scoring

Each question is scored from 0 to 3, with a maximum possible score of 42 across all 14 questions.

Scoring and Interpretation

- **0–14**: High risk – Immediate action required to address significant gaps in AI maturity.

- **15–28**: Moderate risk – Areas for improvement identified, and a structured plan is needed to advance AI maturity.

- **29–42**: Low risk – Robust AI maturity practices in place, but continuous monitoring and minor adjustments are recommended.

Threshold for Passing

Organizations should aim for a minimum score of **28** to ensure they have adequately addressed key AI maturity considerations and are prepared to progress effectively through the maturity model.

CHAPTER 8 DETERMINING AI MATURITY FOR YOUR ORGANIZATION

Summary

As organizations navigate the labyrinth of AI adoption, the maturity model serves as both compass and roadmap – but are you charting your course with precision or stumbling in the dark? From foundational data infrastructure to the dizzying heights of multiagent systems, each level of AI maturity presents its own Everest to climb, demanding not just technological prowess but a holistic transformation of organizational DNA. For boards and C-suites, the imperative is clear: strategically shepherd your company through this digital metamorphosis, balancing the siren call of innovation with the steadfast principles of responsible AI. As you ascend the peaks of AI sophistication, remember that true maturity isn't just about technological advancement – it's about fostering an ecosystem where AI and human ingenuity symbiotically propel your organization toward unprecedented horizons. In this brave new world of silicon-enhanced decision-making, will your organization be a trailblazer or a cautionary tale?

CHAPTER 9

Structuring AI Teams for Success: Models for Scaling AI Operations

Effective organizational structure is crucial for the successful deployment and scaling of AI initiatives. As businesses integrate AI into their operations, they must consider various team structures to optimize performance, collaboration, and innovation. This chapter explores different models for structuring AI teams, examining the pros and cons of each approach. By understanding these models, corporate boards and senior executives can choose the best structure to support their AI strategy and drive business growth.

Models for Structuring AI Teams

Organizations can adopt various models to structure their AI teams, each with distinct advantages and challenges. The primary models include Functional, Centralized, Decentralized, Factory, Center of Excellence (CoE), and Consulting. Each model serves different strategic needs and organizational contexts.

CHAPTER 9   STRUCTURING AI TEAMS FOR SUCCESS: MODELS FOR SCALING AI OPERATIONS

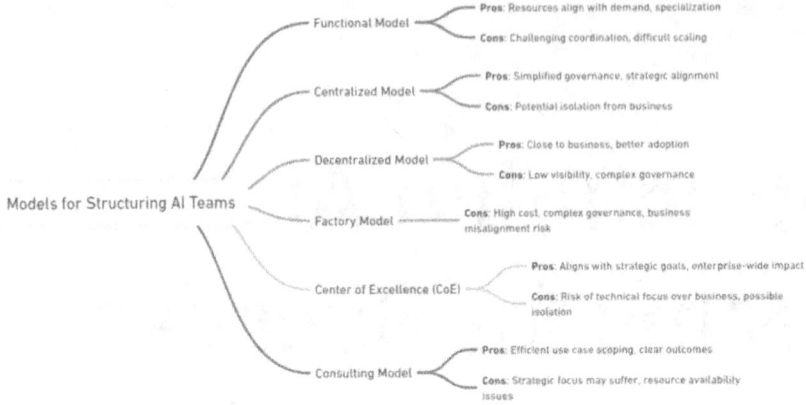

1. Functional Model

In the functional model, AI and analytics teams are dispersed across the organization, with a small central analytics unit.

Pros:

- Resources are where the current demand of the enterprise is.
- Resources can specialize in each functional area.

Cons:

- Coordination is challenging due to team dispersion.
- Lack of a common technical and functional framework.
- Enterprise scaling can be difficult without centralized direction.

2. Centralized Model

A centralized model consolidates AI teams, resources, tools, and data into a single location, accessible by other units.

Pros:

- Easier governance, standards, and management.
- Centralized resources facilitate strategic mission alignment.
- Collaboration and asset sharing are simplified.

Cons:

- Potential lack of co-creation opportunities with business stakeholders.
- Teams may become isolated from business nuances.
- Adoption of AI solutions can be challenging if disconnected from business drivers.

3. Decentralized Model

In this model, AI resources are spread across various functions within the organization, operating independently but aligned with their specific business units.

Pros:

- Better oversight of resources within each function.
- Business skills are closer to teams, enhancing understanding and relevance.
- Solutions are viewed as "business-led," promoting adoption.

Cons:

- Dispersed teams can lead to low visibility of analytics activities.
- Difficult to maintain consistent governance and standards.
- Knowledge sharing and asset management are more complex.

4. Factory Model

The factory model organizes AI teams to prioritize the industrialization of AI and data science solutions, covering the full lifecycle.

Cons:

- Requires significant organization and governance.
- High initial setup cost and ongoing coordination.
- Potential risk of overemphasis on technical execution over business priorities.

5. Center of Excellence (CoE)

A CoE centralizes AI operations and activities, aligning them with the enterprise's strategic goals.

Pros:

- Aligns AI initiatives with strategic business goals.
- Creates more enterprise-wide initiatives.
- Facilitates better stakeholder relationships.

Cons:

- Risk of becoming too focused on technical excellence at the expense of business priorities.
- May turn into a resource pool instead of driving strategic growth.
- Requires careful balance to avoid being seen as isolated from core business operations.

6. Consulting Model

The consulting model leverages AI resources based on project availability and needs, often charging the organization for services.

Pros:

- Efficient scoping of use cases.
- Clear value and outcomes in a structured framework.
- Encourages a "skin in the game" approach.

Cons:

- Strategic initiatives can be overlooked due to service payment focus.
- Availability of resources can become an issue.
- Lack of centralized location requires ongoing marketing efforts to ensure awareness.

Choosing the Right Model

Selecting the appropriate AI team structure depends on various factors, including organizational size, industry, strategic goals, and resource availability. Here are strategic considerations for making this decision:

CHAPTER 9   STRUCTURING AI TEAMS FOR SUCCESS: MODELS FOR SCALING AI OPERATIONS

1. **Align Structure with Strategic Goals**
   The chosen model should align with the organization's broader strategic goals. For example, a decentralized model might suit organizations seeking to embed AI deeply within each business unit, while a centralized model could be better for those prioritizing governance and standardization.

2. **Consider Resource Availability and Specialization**
   Assess the availability and specialization of AI resources. Centralized and factory models can maximize the use of specialized talent, whereas functional and decentralized models might better leverage domain-specific knowledge within business units.

3. **Evaluate Governance and Coordination Needs**
   Strong governance and coordination are crucial for the success of AI initiatives. Centralized and CoE models offer better control and standardization, while decentralized models require robust frameworks to maintain consistency across units.

4. **Foster Collaboration and Innovation**
   Choose a model that promotes collaboration and innovation. Factory and CoE models are conducive to creating synergies and driving large-scale AI projects, while consulting models can bring fresh perspectives and specialized expertise for specific projects.

5. **Balance Technical and Business Priorities**
   Ensure that the AI structure balances technical excellence with business priorities. Factory and centralized models may focus heavily on technical execution, necessitating mechanisms to keep business goals at the forefront.

Strategic Implementation of AI Structures

To effectively implement the chosen AI team structure, organizations should follow these strategic steps:

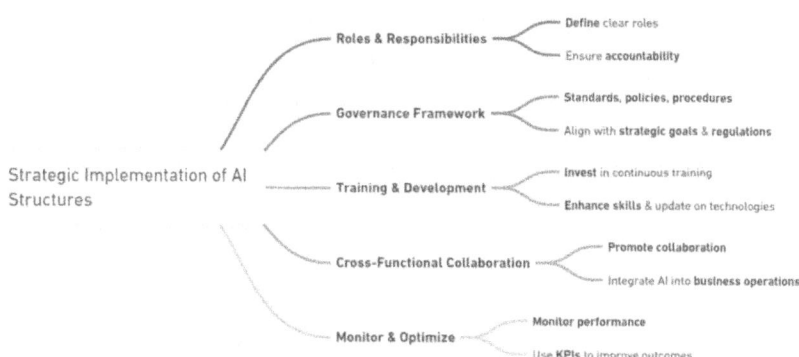

CHAPTER 9 STRUCTURING AI TEAMS FOR SUCCESS: MODELS FOR SCALING AI OPERATIONS

1. **Define Clear Roles and Responsibilities**
   Establish clear roles and responsibilities for AI teams to avoid overlaps and ensure accountability. This clarity helps streamline processes and improves coordination.

2. **Develop a Governance Framework**
   Implement a governance framework that includes standards, policies, and procedures for AI initiatives. This framework should ensure alignment with strategic goals and compliance with regulations.

3. **Invest in Training and Development**
   Provide continuous training and development opportunities for AI teams. This investment enhances skills, keeps teams updated on the latest technologies, and fosters a culture of continuous improvement.

4. **Encourage Cross-Functional Collaboration**
   Promote cross-functional collaboration to integrate AI initiatives seamlessly into business operations. This approach ensures that AI solutions address real business challenges and drive tangible value.

5. **Monitor and Optimize Performance**
   Regularly monitor the performance of AI teams and projects. Use metrics and KPIs to track progress, identify areas for improvement, and optimize processes for better outcomes.

CHAPTER 9 STRUCTURING AI TEAMS FOR SUCCESS: MODELS FOR SCALING AI OPERATIONS

AI Team Structure Success Checklist for Board and C-Suite

Scoring System

- **0 points:** Not addressed
- **1 point:** Partially addressed
- **2 points:** Fully addressed, but needs improvement
- **3 points:** Fully addressed and well-executed

Functional Model

1. **Resource Allocation** (Score: 0-3)
    - Are resources strategically allocated to meet the current demands of the enterprise?
2. **Specialization** (Score: 0-3)
    - Are resources specialized within each functional area?

Centralized Model

3. **Governance and Management** (Score: 0-3)
    - Is there ease of governance, standards, and management in the centralized model?
4. **Resource Consolidation** (Score: 0-3)
    - Are AI teams, resources, tools, and data consolidated in a single location?

Decentralized Model

5. **Functional Alignment** (Score: 0-3)
   - Are AI resources aligned with their specific business units?

6. **Business Integration** (Score: 0-3)
   - Is there a close integration of business skills with AI teams?

Factory Model

7. **Lifecycle Management** (Score: 0-3)
   - Does the model prioritize the industrialization of AI solutions across their lifecycle?

8. **Organizational Coordination** (Score: 0-3)
   - Are there significant organization and governance mechanisms in place?

Center of Excellence (CoE)

9. **Strategic Alignment** (Score: 0-3)
   - Does the CoE align AI initiatives with the enterprise's strategic goals?

10. **Enterprise Initiatives** (Score: 0-3)
    - Are AI operations and activities centralized to facilitate enterprise-wide initiatives?

Consulting Model

11. **Project Scoping** (Score: 0-3)
    - Are AI resources leveraged based on project availability and needs?

12. **Outcome Structuring** (Score: 0-3)
    - Is there a clear structure for value and outcomes from AI projects?

Strategic Considerations

13. **Alignment with Strategic Goals** (Score: 0-3)
    - Does the chosen model align with the organization's broader strategic goals?

14. **Resource Specialization** (Score: 0-3)
    - Does the model maximize the use of specialized talent and domain-specific knowledge?

Governance and Coordination

15. **Governance Framework** (Score: 0-3)
    - Is there a strong governance framework to maintain consistency and standardization?

16. **Coordination Mechanisms** (Score: 0-3)
    - Are there robust coordination mechanisms across units?

Collaboration and Innovation

17. **Promotion of Collaboration** (Score: 0-3)
    - Does the model promote cross-functional collaboration and innovation?

18. **Balance of Technical and Business Priorities** (Score: 0-3)
    - Is there a balance between technical excellence and business priorities?

Strategic Implementation Steps

19. **Clear Roles and Responsibilities** (Score: 0-3)
    - Are clear roles and responsibilities established to avoid overlaps and ensure accountability?

20. **Training and Development** (Score: 0-3)
    - Are continuous training and development opportunities provided for AI teams?

Monitoring and Optimization

21. **Performance Monitoring** (Score: 0-3)
    - Are there metrics and KPIs in place to monitor and optimize AI team performance?

22. **Continuous Improvement** (Score: 0-3)
    - Is there a system for continuous improvement and adaptation of AI processes?

Total Scoring

Each question is scored from 0 to 3, with a maximum possible score of 66 across all 22 questions.

Scoring and Interpretation

- **0–22**: High risk – Immediate action required to address significant gaps in AI team structuring.

- **23–44**: Moderate risk – Areas for improvement identified, and a structured plan is needed to enhance AI team structuring.

- **45–66**: Low risk – Robust AI team structuring practices in place, but continuous monitoring and minor adjustments are recommended.

Threshold for Passing

Organizations will typically choose one of the above-mentioned models. Organizations should aim for a minimum score of **22** to ensure they have adequately addressed key AI team structuring considerations and are prepared to manage and scale AI operations effectively.

Summary

An effective organizational structure is key to deploying and scaling AI initiatives. This chapter examines different models for structuring AI teams, including Functional, Centralized, Decentralized, Factory, Center of Excellence (CoE), and Consulting models—each with its unique advantages and challenges. The choice of model depends on factors such as company size, industry, and strategic priorities.

CHAPTER 9 STRUCTURING AI TEAMS FOR SUCCESS: MODELS FOR SCALING AI OPERATIONS

The chapter provides a guide to selecting the right structure, emphasizing the importance of strategic alignment, resource allocation, governance, and fostering collaboration. Clear roles, governance frameworks, cross-functional cooperation, and ongoing training are highlighted as essential for scaling AI effectively. Ultimately, a well-structured AI team enables organizations to transform AI from isolated projects into drivers of growth, ensuring that AI initiatives align with broader business objectives.

CHAPTER 10

AI Partnerships and Alliances

Collaborating with the right partners can provide access to cutting-edge technologies which your organizations will need years to develop on your own, domain expertise, and a broader ecosystem that facilitates rapid progress. But it's crucial to approach these partnerships with careful consideration because the true power of a business partnership lies not in the avoidance of losses, but in the collective ability to turn challenges into opportunities.

Let us explore the paramount considerations that should guide decision-making when forging AI partnerships.

- **Strategic Alignment:** Ensure that the chosen AI partners align with the company's strategic goals and vision. It is vital to identify partners who share a common purpose and possess complementary capabilities.

  In addition, access the partner's expertise in AI domains, their track record in delivering successful projects, particularly the use cases your organization has prioritized, their experience within your industry and their understanding of your end customer needs.

  This crucial alignment will lay strong foundations for a fruitful and long-lasting collaboration.

- **Technical Proficiency:** Evaluate the technical proficiency and capabilities of potential AI partners.

  Do they have product or platform offerings, or do they excel in advisory or system integration?

  It all depends on your organization's needs and priorities but look for partners who have a deep understanding of AI algorithms, data engineering, model deployment and cloud migration. Assess their ability to handle complex datasets with speed while ensuring data privacy and security and implement robust validation and testing processes. Technical excellence is crucial for successful AI implementation and eventually adoption.

- **Ecosystem and Collaboration:** Consider the partner's ecosystem and collaborative approach. Look for partners who have established relationships with academic institutions, research organizations, and technology providers.

  An extensive network allows for knowledge-sharing, access to the latest research, and the ability to leverage best practices. Additionally, seek partners who are committed to collaboration, co-creation, and knowledge transfer, fostering a strong partnership culture.

- **Change Management and Talent**: It's crucial to evaluate the partner's change management capabilities and their ability to support the organization's talent development.

  Most projects fail to deliver desired business outcomes not because they never got delivered, it's because the delivered programs or products never got fully adopted.

- **AI adoption requires cultural and organizational shifts.** Look for partners who have expertise in change management and can support the change management process, helping employees adapt to new ways of working. Another aspect to consider is to assess the partner's ability to attract and develop top AI talent, as access to skilled talent is crucial for successful implementation.

- **Ethical Considerations:** Pay close attention to ethical considerations when selecting AI partners. Ensure they adhere to ethical guidelines and practices in data usage, algorithm transparency, and bias mitigation. Consider partners who prioritize responsible AI practices, inclusivity, and fairness. Ethical alignment will help maintain the company's reputation and build trust with customers, employees, and stakeholders.

- **Risk Management and Governance:** Assessing partner's risk management and governance practices is important. Evaluating their ability to handle risks related to data privacy, cybersecurity, regulatory compliance, and intellectual property can save your organization costly surprises. Seek partners who have robust governance frameworks in place to ensure transparency, accountability, and risk mitigation throughout the partnership.

AI Partnerships and Alliances Checklist for Board and C-Suite
Scoring System

- **0 points:** Not addressed
- **1 point:** Partially addressed
- **2 points:** Fully addressed, but needs improvement
- **3 points:** Fully addressed and well-executed

Checklist Questions
Strategic Alignment

1. **Alignment with Strategic Goals** (Score: 0-3)
   - Does the AI partner align with the company's strategic goals and vision?

2. **Complementary Capabilities** (Score: 0-3)
   - Does the partner possess complementary capabilities that add value to the collaboration?

3. **Industry Experience and Understanding** (Score: 0-3)
   - Does the partner have a proven track record in your industry and an understanding of your end customer needs?

4. **Use Case Relevance** (Score: 0-3)
   - Has the partner successfully delivered projects in use cases similar to those prioritized by your organization?

Technical Proficiency

5. **Technical Expertise** (Score: 0-3)
   - Does the partner have deep expertise in AI algorithms, data engineering, model deployment, and cloud migration?

6. **Data Handling and Privacy** (Score: 0-3)
   - Can the partner handle complex datasets efficiently while ensuring data privacy and security?
7. **Validation and Testing** (Score: 0-3)
   - Does the partner implement robust validation and testing processes to ensure the quality and reliability of AI models?

Ecosystem and Collaboration

8. **Ecosystem Integration** (Score: 0-3)
   - Does the partner have established relationships with academic institutions, research organizations, and technology providers?
9. **Collaborative Culture** (Score: 0-3)
   - Is the partner committed to collaboration, co-creation, and knowledge transfer, fostering a strong partnership culture?

Change Management and Talent

10. **Change Management Expertise** (Score: 0-3)
    - Does the partner have expertise in change management to support the cultural and organizational shifts required for AI adoption?
11. **Talent Development** (Score: 0-3)
    - Can the partner attract and develop top AI talent to support the successful implementation of AI initiatives?

Ethical Considerations

12. **Ethical Guidelines Adherence** (Score: 0-3)
    - Does the partner adhere to ethical guidelines and practices in data usage, algorithm transparency, and bias mitigation?

13. **Responsible AI Practices** (Score: 0-3)
    - Does the partner prioritize responsible AI practices, inclusivity, and fairness in their AI development and deployment?

Risk Management and Governance

14. **Risk Management Capabilities** (Score: 0-3)
    - Can the partner effectively manage risks related to data privacy, cybersecurity, regulatory compliance, and intellectual property?

15. **Governance Framework** (Score: 0-3)
    - Does the partner have a robust governance framework in place to ensure transparency, accountability, and risk mitigation throughout the partnership?

Total Scoring

Each question is scored from 0 to 3, with a maximum possible score of 45 across all 15 questions.

Scoring and Interpretation

- **0–15**: High risk – Immediate action required to address significant gaps in AI partnership considerations.

- **16–30**: Moderate risk – Areas for improvement identified, and a structured plan is needed to enhance AI partnerships.

- **31–45**: Low risk – Strong AI partnership considerations in place, but continuous monitoring and minor adjustments are recommended.

Threshold for Passing

Organizations should aim for a minimum score of **30** to ensure they have adequately addressed key considerations for AI partnerships and alliances. This score indicates a solid foundation for collaboration, mitigating risks while maximizing opportunities for success.

Summary

In summary, partnerships are not just beneficial but sometimes needed to accelerate you on a journey. But choose your business partners wisely since a good business partnership not only mitigates risks but actively seeks opportunities, transforming potential losses into valuable lessons and wins. A misguided alliance, however, can transform a promising venture into new challenges, conflicts, and missed opportunities.

CHAPTER 11

AI Budgets and Investments

Artificial intelligence (AI) is rapidly becoming a mainstream technology, with businesses of all sizes investing in its potential to improve their operations and bottom line and the proof is in the pudding. More than, 60% of organizations now use AI in some capacity, up from 35% in 2017.

This trend is being driven by several factors, including the increasing availability of data, the falling cost of computing power, and the development of more sophisticated AI algorithms. As a result, AI is being used to automate tasks, improve decision-making, and create new products and services.

The Trends in AI Investment in the Organizations

One long-standing investment area has been increased focus on AI-powered automation. Businesses are using AI to automate tasks that were previously done by humans, such as customer service, fraud detection, and medical diagnosis, leading to significant cost savings and productivity gains.

CHAPTER 11 AI BUDGETS AND INVESTMENTS

Another AI adoption trend is the growing use of AI for decision-making. AI is being used to make better decisions about everything from product development to marketing to risk management. This is helping businesses to improve their efficiency, effectiveness, and profitability. Although AI system can detect pattern effectively, they still cannot do logical reasoning effectively, so using them to make mission-critical decisions will be a losing proposition in the long run.

Finally, there is a growing interest in the use of AI to create new products and services. AI is being used to develop new drugs, create new forms of entertainment, and design new ways of working. This is leading to a revolution in the way businesses operate and the way people live.

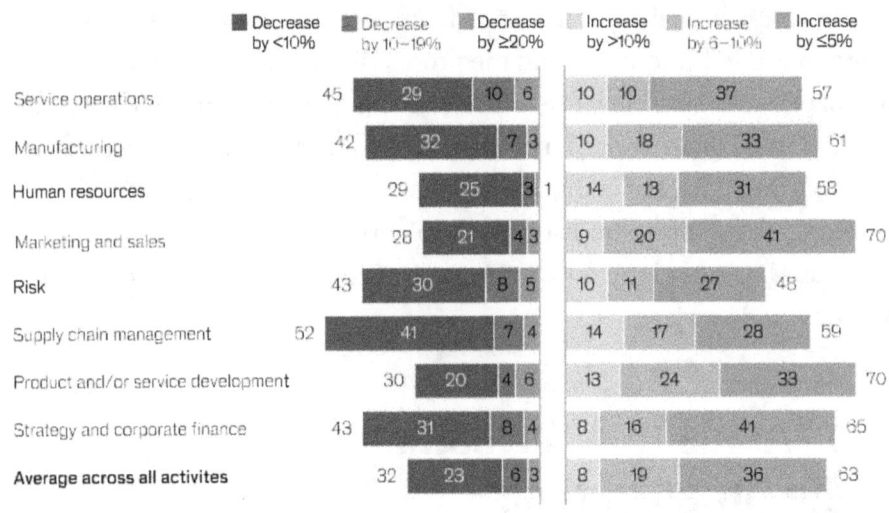

Source: *McKinsey & Company*

CHAPTER 11　AI BUDGETS AND INVESTMENTS

AI Budgets and Investments: Checklist for Board and C-Suite

Scoring System

- **0 points:** Not addressed
- **1 point:** Partially addressed
- **2 points:** Fully addressed, but needs improvement
- **3 points:** Fully addressed and well-executed

Checklist for AI Budgets and Investments

1. **Strategic Alignment with Business Goals** (Score: 0-3)
   - Is the AI investment strategy aligned with the organization's overall business goals?
   - Are AI investments prioritized based on their potential to drive revenue growth, cost savings, or competitive advantage?

2. **Budget Allocation for AI Projects** (Score: 0-3)
   - Is there a dedicated budget for AI initiatives that is proportional to their strategic importance?
   - Are funds allocated effectively across various AI projects, including research, development, and deployment?

3. **ROI Expectations and Monitoring** (Score: 0-3)
   - Are clear ROI expectations set for AI investments?
   - Is there a mechanism to monitor and measure the actual ROI of AI projects over time?

4. **Investment in AI Infrastructure** (Score: 0-3)
   - Are sufficient resources allocated to building and maintaining the technological infrastructure necessary for AI, such as computing power, data storage, and networking?
   - Does the infrastructure investment support the scalability and long-term sustainability of AI initiatives?

5. **AI Talent Acquisition and Development** (Score: 0-3)
   - Is there an adequate budget for attracting, retaining, and developing AI talent, including data scientists, machine learning engineers, and AI specialists?
   - Are ongoing training and development programs funded to keep the AI team's skills up to date?

6. **Risk Management and Compliance** (Score: 0-3)
   - Are AI investments made with a clear understanding of associated risks, including regulatory compliance, ethical considerations, and data security?
   - Is there a budget allocated for mitigating these risks, such as through legal consultation, security measures, and compliance audits?

7. **Adoption and Integration Costs** (Score: 0-3)

    - Are the costs of adopting and integrating AI solutions into existing workflows and systems accounted for in the budget?

    - Is there financial support for change management initiatives to ensure smooth AI adoption across the organization?

8. **AI Research and Innovation** (Score: 0-3)

    - Does the budget include funding for AI research and innovation, including exploring new technologies and methodologies?

    - Are there investments in pilot projects or experimental AI applications that could lead to breakthrough innovations?

9. **Long-Term Investment Planning** (Score: 0-3)

    - Is there a long-term investment plan for AI that considers future technological advancements and the evolving business landscape?

    - Does the plan include provisions for scaling successful AI initiatives and decommissioning projects that do not deliver expected value?

10. **Evaluation of AI Investment Partners** (Score: 0-3)

    - Are potential AI vendors and partners thoroughly evaluated for their ability to deliver on investment expectations?

    - Is there a process in place to ensure that partnerships are cost-effective and strategically beneficial?

11. **Impact on Topline and Bottomline** (Score: 0-3)
    - Are AI investments targeted at not only cost savings but also at generating new revenue streams and enhancing customer experiences?
    - Is there a focus on leveraging AI to create new products, services, or business models that impact the topline?

12. **Sustainability and Ethical Considerations** (Score: 0-3)
    - Are sustainability and ethical considerations factored into AI investment decisions?
    - Does the budget include initiatives to ensure that AI deployment is environmentally sustainable and socially responsible?

Total Scoring

Each question is scored from 0 to 3, with a maximum possible score of 36 across all 12 questions.

Scoring and Interpretation

- **0-12**: High risk – Immediate action required to address significant gaps in AI budget planning and investment strategy.
- **13-24**: Moderate risk – Areas for improvement identified, and a structured plan is needed to enhance AI investment effectiveness.

- **25–36**: Low risk – Robust AI budget and investment strategies in place, but continuous monitoring and minor adjustments are recommended.

Threshold for Passing

Organizations should aim for a minimum score of **25** to ensure they have adequately addressed key AI budget and investment considerations and are prepared to manage and scale AI operations effectively.

Summary

As AI permeates more into organizations, it will start to take more value-added tasks like enhancing customer service or helping commercial team upsell a product. And when AI start impacting topline rather than just a cost savings play, that's when the C-suites and boards will have a burning interest in its adoption across their organizations.

CHAPTER 12

AI Change Management

Integrating AI into your organization involves significant changes, which can be met with excitement by some and skepticism by others. This is where change management becomes vital. It involves crafting a strategic plan to ensure that new AI solutions are not just implemented but are fully embraced and integrated into the daily workflows of employees.

Tip 1: Set Ambitious Goals, Start with Small Steps

While setting ambitious goals for AI adoption is important, implementing AI changes without a clear vision can be counterproductive. To align AI success with organizational objectives, it is essential to establish a clear vision and specific business goals for AI capabilities. Starting with small, manageable projects allows for demonstrating tangible value, building trust, and setting the stage for broader AI initiatives.

Tip 2: Prioritize Human-Centered Design

Incorporating end-users into every phase of AI system development – from concept to testing – ensures that technology solutions meet the actual needs of those who will use them. This approach underscores the importance of considering technology as an integral part of the change management plan.

Case Study: Enhancing Emergency Response Times

Effective collaboration between subject matter experts, data scientists, and AI professionals can yield AI solutions that empower better human decision-making. For example, in 2019, Fairfax County, Virginia, aimed to reduce emergency medical response times to under four minutes. Through the strategic application of AI, they improved performance despite challenges such as fixed station locations and a growing population. This success highlights the importance of following the strategies outlined in this chapter.

Cost Efficiency and Growth: The Dual Benefits of AI

AI systems enhance operational efficiency and productivity, leading to significant cost savings and quality improvements. At the same time, AI drives growth by enabling the development of innovative products, services, and customer experiences. By automating routine tasks, AI frees employees to focus on higher-value activities, reducing operational costs and enhancing job satisfaction. Furthermore, AI's data analysis capabilities help identify inefficiencies and optimize processes, providing additional cost control.

OCM: Emphasizing the Human Aspect of AI Change Management

A people-centric approach is essential for successful AI integration. Effective AI change management programs consider the human dimensions of AI implementation, such as job role reconfigurations and fostering an organizational culture that embraces change. Involving employees in the AI implementation process demystifies the technology and showcases its potential to enhance their work rather than threaten their job security. This approach leads to higher levels of employee engagement and acceptance, which are critical for successful transformation.

Leadership's Role in AI-Driven Change

Leadership is critical in supporting AI-driven change. Leaders must advocate for the change, demonstrate commitment, and lead by example. Using language that emphasizes augmentation rather than substitution helps alleviate employee fears. Exemplary leadership involves providing the necessary resources and support for the AI change management process, including technical tools, training, and support structures. Building coalitions and celebrating incremental successes further reinforces leadership support.

Addressing Resistance to Change

Resistance to change is a common challenge during AI adoption. Employees may fear that AI will make their jobs obsolete or that they lack the necessary skills for an AI-driven environment. Applying organizational change management (OCM) to AI adoption can address these fears by providing a clear vision of the change, its necessity, and its benefits for both the organization and its employees.

CHAPTER 12   AI CHANGE MANAGEMENT

Effective Communication and Training

Clear, consistent, and transparent communication is vital for managing expectations, alleviating fears, and building trust. Communication should explain the reasons for AI adoption, its benefits, and its impact on various stakeholders. Encouraging questions and feedback creates an environment of trust and openness. Training equips employees with the necessary skills to operate in an AI-driven environment, bridging the gap between current capabilities and future requirements. Successful AI training programs foster a culture of continuous learning and development, ensuring the workforce remains agile and prepared for future technological advancements.

PPM: The Structural Backbone of AI Change Management

Effective portfolio and program management (PPM) plays a crucial role in AI adoption, ensuring alignment with business goals, efficient resource allocation, and successful execution. Portfolio management oversees all AI-related initiatives within the organization, prioritizing them based on potential return on investment, feasibility, and alignment with business objectives. Program management coordinates related projects to achieve benefits that individual projects cannot achieve in isolation, ensuring successful delivery and execution of AI initiatives.

PPM's Contribution to Successful AI Adoption

PPM provides the necessary oversight, coordination, and control to ensure AI initiatives deliver expected benefits and align with strategic objectives. By fostering a disciplined approach to AI adoption, PPM ensures that AI initiatives are pursued as part of a coherent strategy rather than in isolation, avoiding wasted resources and misalignment.

CHAPTER 12   AI CHANGE MANAGEMENT

Strategic Considerations for AI Adoption: Ensuring Effective Implementation

Key Considerations for AI Adoption

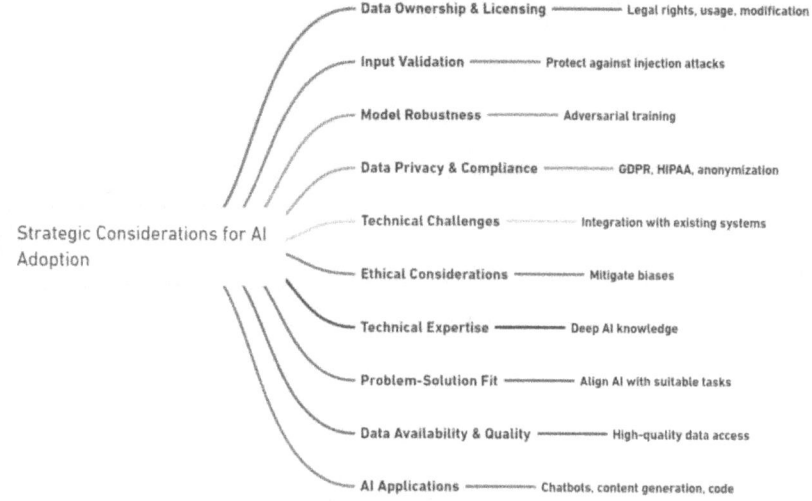

1. Data Ownership and Licensing

Ensuring legal rights to use training data is fundamental for AI projects. Organizations must understand data ownership implications, including usage, modification, and resale rights. This involves verifying whether proprietary data used in training AI models remains with the company or the provider.

Example: An Independent Software Vendor (ISV) must check if they retain ownership of proprietary data used in training their model or if it remains with the provider.

139

2. Input Validation and Sanitization

Protecting AI systems from injection attacks requires robust validation and sanitization of inputs. Implementing checks to filter out harmful data inputs is essential to prevent manipulation of the system.

Example: Implementing checks to filter out harmful data inputs that could manipulate the system.

3. Model Robustness

Increasing resistance to adversarial attacks involves adversarial training and input filtering. Training models with a variety of challenging inputs improves their ability to handle real-world adversarial scenarios.

Example: Training a model with a variety of challenging inputs to improve its ability to handle real-world adversarial scenarios.

4. Data Privacy and Compliance

Adhering to data protection laws such as GDPR or HIPAA is crucial for building user trust and compliance. This includes measures like anonymization and encryption to protect user data.

Example: Encrypting user data and ensuring all processing complies with relevant privacy regulations.

5. Technical Challenges

Addressing integration with existing systems and ensuring AI implementations enhance workflows is critical for smooth adoption. Developing APIs that allow seamless integration with key business systems helps avoid disruptions.

Example: Developing APIs that allow seamless integration of AI with key business systems without disrupting existing workflows.

6. Ethical Considerations

Mitigating biases in AI applications is essential for ethical AI deployment. This involves diversifying training data and adhering to ethical guidelines to avoid biased outcomes.

Example: Training AI models on diverse datasets to avoid biased outcomes in AI-generated content or decisions.

7. Technical Expertise

Requiring deep technical knowledge for sustainable AI deployment and maintenance is non-negotiable. Building a team with expertise in machine learning, data science, and AI lifecycle management is crucial.

Example: Building a team with skills in machine learning, data science, and AI lifecycle management.

8. Problem–Solution Fit

Aligning AI capabilities with the right problems ensures effective technology application. This involves using AI where it is most suitable, rather than for tasks requiring deep reasoning.

Example: Using generative AI for summarizing texts where it is most suitable, rather than for tasks requiring deep reasoning.

9. Data Availability and Quality

Ensuring access to high-quality and relevant data is vital for effective training and outputs. Maintaining a large, clean dataset of customer interactions, for example, is essential for training AI systems.

Example: Maintaining a large, clean dataset of customer interactions for training a customer service chatbot.

CHAPTER 12   AI CHANGE MANAGEMENT

10. Common AI Applications

Highlighting applications across industries such as chatbots, content generation, code generation, data augmentation, and search improvements shows the versatility of AI. Leveraging AI in these areas can significantly enhance business operations.

Example: Using AI to create chatbots for customer support, generate marketing copy, or assist in coding.

Strategic Directions for AI Adoption

To ensure successful AI adoption, organizations should consider the following strategic directions:

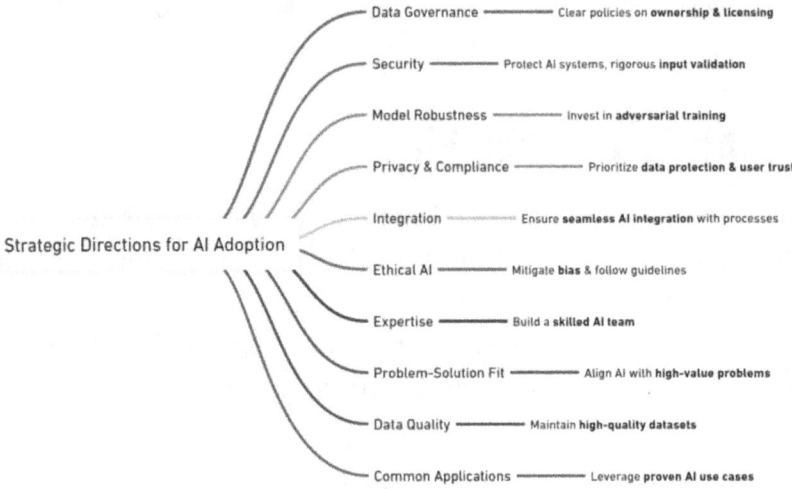

1. **Establish Clear Data Governance**

    Developing clear policies for data ownership and licensing is essential. Organizations must ensure they have legal rights to all data used in AI training and operations. This includes understanding the implications of data ownership on usage, modification, and resale.

CHAPTER 12   AI CHANGE MANAGEMENT

2. **Implement Robust Security Measures**

   AI systems must be protected against injection attacks through rigorous input validation and sanitization. Implementing comprehensive security protocols ensures the integrity and reliability of AI outputs.

3. **Enhance Model Robustness**

   Investing in adversarial training and input filtering techniques enhances the robustness of AI models. These practices help models withstand adversarial attacks and perform reliably in real-world scenarios.

4. **Prioritize Data Privacy and Compliance**

   Adhering to data protection laws and implementing privacy measures such as anonymization and encryption is critical. Organizations must prioritize user trust and compliance to maintain their reputation and avoid regulatory penalties.

5. **Address Technical Integration Challenges**

   Ensuring seamless integration of AI systems with existing business processes is crucial. Developing APIs and other integration tools helps AI applications enhance, rather than disrupt, existing workflows.

6. **Promote Ethical AI Practices**

   Mitigating biases and adhering to ethical guidelines is essential for responsible AI deployment. Organizations should commit to training AI models on diverse datasets and implementing checks to ensure unbiased outcomes.

7. **Build Technical Expertise**

   Investing in a skilled AI team is fundamental for sustainable AI deployment. Organizations must ensure they have the necessary technical expertise in machine learning, data science, and AI lifecycle management.

8. **Ensure Problem-Solution Fit**

   Aligning AI capabilities with appropriate problems ensures effective application. Organizations should evaluate where AI can add the most value and focus on those areas for deployment.

9. **Maintain High Data Quality**

   Ensuring access to high-quality, relevant data is critical for effective AI training and operation. Organizations must invest in maintaining clean, comprehensive datasets to support their AI initiatives.

10. **Leverage Common AI Applications**

    Identifying and leveraging common AI applications across industries can drive significant business value. By focusing on proven use cases such as chatbots, content generation, and data augmentation, organizations can enhance their operations and customer experiences.

AI Change Management Checklist for Board and C-Suite

Scoring System

- **0 points:** Not addressed
- **1 point:** Partially addressed
- **2 points:** Fully addressed, but needs improvement
- **3 points:** Fully addressed and well-executed

General Compliance

1. **Lawfulness and Transparency** (Score: 0-3 each)
   - Are our AI data processing activities compliant with relevant laws?
   - Is information about data usage communicated transparently to all stakeholders?

Change Management Framework

2. **Clear Vision and Objectives** (Score: 0-3)
   - Have we defined a clear vision and objectives for AI change management?

3. **Alignment with Business Goals** (Score: 0-3)
   - Are the AI change initiatives aligned with the broader business objectives?

CHAPTER 12   AI CHANGE MANAGEMENT

Stakeholder Engagement

4. **Stakeholder Identification** (Score: 0-3)
   - Have all relevant stakeholders been identified and their needs assessed?

5. **Communication Plan** (Score: 0-3)
   - Is there a comprehensive communication plan to keep stakeholders informed and engaged?

Governance and Leadership

6. **Leadership Commitment** (Score: 0-3)
   - Is there strong leadership commitment to AI change management initiatives?

7. **Governance Framework** (Score: 0-3)
   - Have we established a governance framework to oversee AI change management?

Resource Allocation

8. **Adequate Resources** (Score: 0-3)
   - Are sufficient resources allocated for AI change management initiatives?

9. **Resource Utilization** (Score: 0-3)
   - Are resources being utilized efficiently and effectively?

Training and Development

10. **Skills Assessment** (Score: 0-3)
    - Have we conducted a skills assessment to identify training needs?

11. **Training Programs** (Score: 0-3)
    - Are there continuous training and development programs in place for employees?

Monitoring and Evaluation

12. **Performance Metrics** (Score: 0-3)
    - Have we established metrics to monitor the performance of AI change initiatives?

13. **Continuous Improvement** (Score: 0-3)
    - Is there a system for continuous monitoring and improvement of AI change management processes?

Risk Management

14. **Risk Assessment** (Score: 0-3)
    - Have we conducted a risk assessment to identify potential risks associated with AI change initiatives?

15. **Risk Mitigation** (Score: 0-3)
    - Are there strategies in place to mitigate identified risks?

CHAPTER 12 AI CHANGE MANAGEMENT

Cultural Adaptation

16. **Cultural Readiness** (Score: 0-3)

    - Is the organization culturally ready for AI change?

17. **Change Adaptability** (Score: 0-3)

    - Are there measures in place to enhance the organization's adaptability to AI change?

Financial Management

18. **Budget Allocation** (Score: 0-3)

    - Is there a clear budget allocation for AI change management?

19. **Financial Tracking** (Score: 0-3)

    - Are financial expenditures tracked and reported regularly?

Technology Integration

20. **Technology Readiness** (Score: 0-3)

    - Is the current technology infrastructure ready for AI integration?

21. **Integration Plan** (Score: 0-3)

    - Is there a detailed plan for integrating AI technologies into existing systems?

CHAPTER 12   AI CHANGE MANAGEMENT

Measuring Success

22. **ROI Calculation** (Score: 0-3)

    - Do we have a method for calculating the ROI of AI change initiatives?

23. **Impact Analysis** (Score: 0-3)

    - Is there a framework for analyzing the impact of AI changes on business performance?

Total Scoring

Each question is scored from 0 to 3, with a maximum possible score of 69 across all 23 questions.

Scoring and Interpretation

- **0–23**: High risk – Immediate action required to address significant gaps in AI change management.

- **24–46**: Moderate risk – Areas for improvement identified, and a structured plan is needed to enhance AI change management.

- **47–69**: Low risk – Robust AI change management practices in place, but continuous monitoring and minor adjustments are recommended.

CHAPTER 12   AI CHANGE MANAGEMENT

Threshold for Passing

Organizations should aim for a minimum score of **47** to ensure they have adequately addressed key AI change management considerations and are prepared to manage and scale AI operations effectively.

Summary

AI presents organizations with powerful tools to control costs and drive growth. However, to fully realize these benefits, organizations must adopt a people-centric approach to AI integration and leverage strategic portfolio and program management processes. By doing so, organizations can ensure their AI initiatives deliver maximum value and drive long-term, sustainable success. Future discussions will explore how business architecture, process improvement, and performance measurement contribute to successful AI implementation.

CHAPTER 13

AI KPIs and OKRs: Measuring Success and Maximizing Impact

In the rapidly evolving landscape of artificial intelligence (AI), accurately measuring success is not just beneficial – it is imperative. As organizations invest heavily in AI technologies, establishing clear and quantifiable measures of effectiveness ensures that these initiatives are not merely innovative experiments but strategic investments aligned with core business objectives. Utilizing precise metrics and key performance indicators (KPIs) is essential to validate the impact of AI, guide future enhancements, and justify continued or increased investment in these technologies.

The Importance of Metrics in AI Initiatives

Measuring the success of AI initiatives is critical for several reasons.

CHAPTER 13   AI KPIS AND OKRS: MEASURING SUCCESS AND MAXIMIZING IMPACT

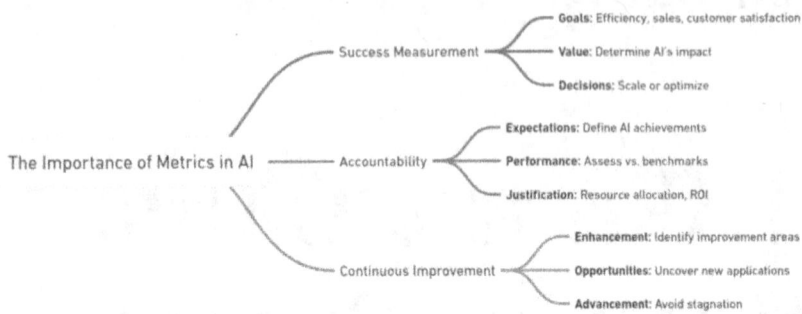

Firstly, it allows organizations to determine whether AI implementations are meeting their intended goals, such as improving efficiency, increasing sales, or enhancing customer satisfaction. Without concrete metrics, it is challenging to ascertain the value AI is adding and difficult to make informed decisions about scaling up or optimizing AI systems.

Secondly, well-defined metrics provide a framework for accountability. They help set expectations for what AI should achieve and offer a means to assess performance against these benchmarks. This is particularly important in justifying the allocation of resources toward AI projects, as stakeholders want to see a return on their investment.

Lastly, metrics facilitate continuous improvement. By regularly measuring the performance of AI systems, companies can identify areas for enhancement, uncover new opportunities for application, and avoid stagnation in a technology landscape that is constantly advancing.

Setting the Context with KPIs and Metrics

Key Performance Indicators (KPIs) and other relevant metrics serve as the cornerstone for gauging AI effectiveness across various business functions. KPIs are quantifiable measurements that reflect the critical success factors of an AI initiative. They might include operational metrics such as speed and accuracy of task completion, financial metrics like ROI and cost savings, or customer-related metrics such as satisfaction scores and engagement rates.

CHAPTER 13   AI KPIS AND OKRS: MEASURING SUCCESS AND MAXIMIZING IMPACT

Choosing the right KPIs involves understanding the specific objectives of each AI project and aligning those objectives with broader business goals. It is not just about tracking any data available but about tracking the right data that provides meaningful insights into how well the AI applications are performing.

Types of AI Metrics

To effectively measure the success of AI initiatives, it is crucial to understand the different types of metrics that can be employed. These metrics can broadly be categorized into efficiency, accuracy, performance, and financial impact. Each category serves a unique purpose and provides insights into various aspects of AI effectiveness.

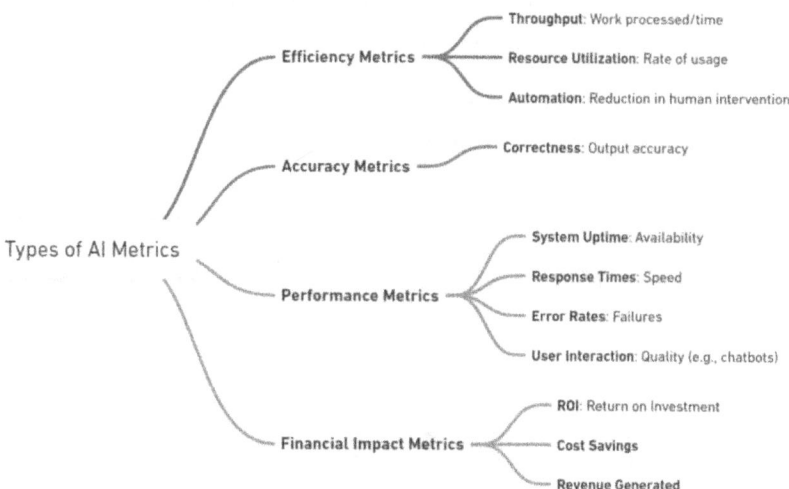

- Efficiency Metrics: These metrics assess how AI technologies streamline operations and reduce the time and resources required to complete tasks. Common efficiency metrics include throughput (the amount of work processed in a given time), resource utilization rates, and the reduction in human intervention for automated processes.

- Accuracy Metrics: Particularly vital in projects involving data processing, prediction, and categorization, accuracy metrics measure the correctness of outputs provided by AI systems. For instance, in a machine learning model used for credit scoring, accuracy metrics would evaluate the percentage of predictions that correctly assessed the creditworthiness of applicants.

- Performance Metrics: These encompass a broader set of indicators that assess the overall effectiveness of AI applications. Performance metrics might include system uptime, response times, error rates, and the quality of user interactions with AI systems, such as chatbots or virtual assistants.

- Financial Impact Metrics: These metrics quantify the economic benefits derived from AI initiatives. They include return on investment (ROI), cost savings, revenue generated from AI-enhanced products or services, and overall financial contributions to business operations.

Selecting the Right Metrics

Choosing the appropriate metrics for an AI project requires a clear understanding of the project's objectives and the specific challenges it aims to address. Here's how to select the right metrics:

CHAPTER 13   AI KPIS AND OKRS: MEASURING SUCCESS AND MAXIMIZING IMPACT

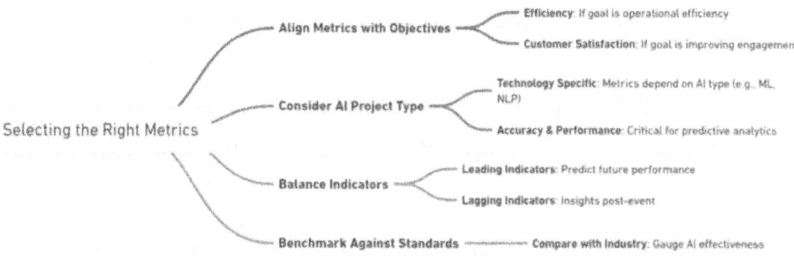

- **Align Metrics with Business Objectives**: Start by clearly defining what the AI project is supposed to achieve. If the goal is to enhance operational efficiency, then efficiency metrics will be most relevant. If improving customer satisfaction is the aim, then metrics related to customer engagement and satisfaction should be prioritized.

- **Consider the Nature of the AI Project**: The type of AI technology implemented (e.g., machine learning models, natural language processing applications) will also influence which metrics are most appropriate. For instance, projects involving predictive analytics will require a strong emphasis on accuracy and performance metrics.

- **Balance Leading and Lagging Indicators**: Include both leading indicators (which predict future performance) and lagging indicators (which provide insights after an event has occurred) to get a comprehensive view of AI effectiveness. For example, the number of AI-driven interactions can be a leading indicator for customer engagement, while customer satisfaction scores are a lagging indicator.

CHAPTER 13   AI KPIS AND OKRS: MEASURING SUCCESS AND MAXIMIZING IMPACT

- **Benchmark Against Industry Standards**: Where possible, compare chosen metrics against industry standards or benchmarks to gauge how your AI initiatives stack up against competitors and industry best practices.

By carefully selecting and applying these metrics, organizations can not only track the immediate impacts of their AI initiatives but also gain insights into how to enhance their long-term strategy. This approach ensures that AI deployments are not just technologically successful but also align closely with broader business goals, driving meaningful improvements across the organization.

Essential KPIs for AI Projects

To effectively gauge the success and impact of AI initiatives, it is crucial to track a set of well-defined Key Performance Indicators (KPIs). These KPIs should cover various aspects of business operations, from operational efficiency to customer satisfaction and revenue growth.

Here's a breakdown of essential KPIs that can help organizations measure the efficacy of their AI projects across these critical areas.

CHAPTER 13   AI KPIS AND OKRS: MEASURING SUCCESS AND MAXIMIZING IMPACT

Operational Efficiency

Improving operational efficiency is often a primary goal of AI projects. The right KPIs can help quantify how AI is enhancing business processes:

- **Process Times**: This KPI measures the time taken to complete specific operations or processes before and after AI integration. A reduction in process times can indicate increased efficiency due to AI automation.

- **Error Rates**: AI is often implemented to reduce human errors in processes such as data entry, calculations, and transaction processing. Tracking error rates before and after AI implementation can highlight improvements in accuracy.

- **Automation Levels**: This metric quantifies the extent of tasks or processes automated by AI, typically expressed as a percentage of total operations. High automation levels can significantly reduce the workload on human employees, allowing them to focus on more strategic tasks.

Customer Satisfaction

AI-enhanced tools are increasingly used in customer service to improve engagement and satisfaction. Relevant KPIs include

- **Response Times**: In customer service, faster response times can lead to higher customer satisfaction. AI can help minimize these times by automating responses and aiding human agents with quicker data retrieval and decision support.

- **Service Quality**: Measured through customer surveys, service quality KPIs assess the effectiveness of AI tools in meeting customer needs. High-quality service is often reflected in positive feedback and low complaint rates.

- **Customer Retention Rates**: Retention rates can be significantly impacted by the quality of service. AI's ability to provide personalized experiences and proactive service can enhance customer loyalty and retention.

Revenue Growth

AI initiatives can also drive revenue growth by enhancing sales processes and marketing efforts. Key metrics in this area include

- **New Leads Generated**: AI tools can help identify and nurture potential leads more effectively than traditional methods. Tracking the number of new leads generated through AI-powered campaigns can demonstrate the impact of AI on sales funnel expansion.

- **Upsell Rates**: AI's capability to analyze customer data and predict buying patterns can be used to increase upsell opportunities. Monitoring upsell rates before and after deploying AI can show how effectively AI contributes to sales growth.

- **Contribution to Sales**: This KPI measures the percentage of total sales directly attributable to AI initiatives, such as recommendations made by AI systems or automated marketing campaigns. It helps quantify the direct financial benefits of AI to the business.

CHAPTER 13   AI KPIS AND OKRS: MEASURING SUCCESS AND MAXIMIZING IMPACT

Quantifying AI's ROI

For business leaders, the decision to invest in AI hinges significantly on the ability to demonstrate a clear return on investment (ROI). Quantifying the ROI of AI initiatives involves a comprehensive analysis of both the tangible and intangible benefits AI brings to an organization, balanced against the costs incurred during its implementation. This section outlines how to effectively calculate and present AI's ROI, ensuring that business leaders can make informed decisions based on solid financial and strategic insights.

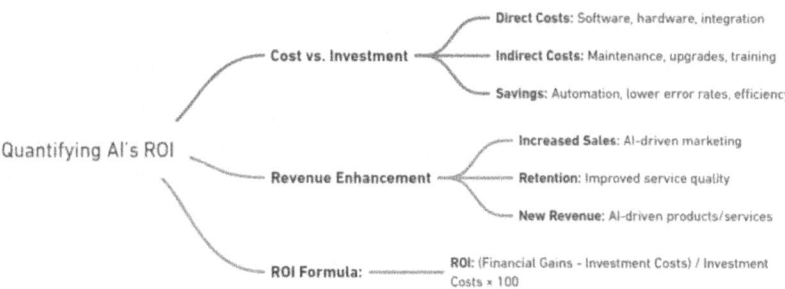

Cost Savings vs. Investment Costs

Calculating the ROI of AI projects requires a detailed assessment of both the direct and indirect costs and savings associated with these initiatives:

- **Direct and Indirect Costs**: Start by compiling all costs related to the development and deployment of AI systems. This includes upfront expenses like software acquisition, hardware investments, and costs of integrating AI into existing systems. Additionally, consider ongoing costs such as maintenance, upgrades, training employees to use AI tools, and potential increases in operational costs due to new technology.

- **Operational Cost Savings:** AI often leads to significant reductions in operational costs. This can be quantified by measuring the decrease in labor costs due to automation, lower error rates leading to fewer reworks, and enhanced efficiency in processes that reduce resource consumption. Collect data before and after AI implementation to clearly demonstrate these savings.

- **Revenue Enhancements:** Besides cost savings, AI can drive revenue growth through various channels. Increased sales from AI-enhanced marketing, higher customer retention rates due to improved service quality, and new revenue streams from AI-driven products or services are critical to consider. These should be quantified to the extent possible and included in the ROI calculation.

- **Calculating ROI**: The ROI can be calculated using the formula:

  $$ROI = \frac{(\text{Financial Gains} - \text{Investment Costs})}{\text{Investment Costs}} \times 100$$

  This formula provides a percentage that reflects the return on every dollar invested in AI, considering both savings and additional revenue.

Challenges in Measuring AI Success

Accurately measuring the success of AI initiatives is critical for validating their impact and guiding future investments.

CHAPTER 13   AI KPIS AND OKRS: MEASURING SUCCESS AND MAXIMIZING IMPACT

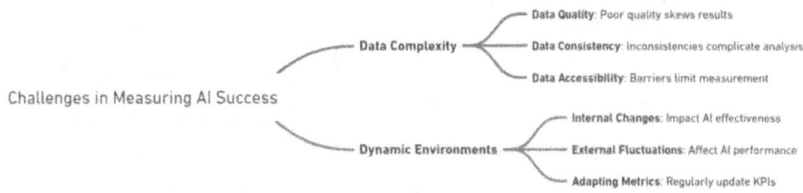

However, this task is often complicated by several inherent challenges, particularly related to data management and the dynamic nature of business environments. Addressing these challenges is essential to ensure that AI metrics and KPIs provide reliable and actionable insights.

- **Data Complexity**: One of the primary obstacles in measuring AI success is the complexity of data involved. AI systems require large volumes of data to train and operate, and the quality of this data can significantly affect the outcomes of AI applications.

- **Data Quality**: Poor data quality can skew AI performance and lead to inaccurate conclusions about its effectiveness. Inconsistencies, incomplete data sets, and erroneous entries all undermine the integrity of AI outputs. Ensuring data cleanliness and accuracy is paramount for reliable measurement.

- **Data Consistency**: AI systems often pull data from multiple sources, which may not always be synchronized or standardized. Differences in data formats or updates can lead to inconsistencies that complicate analysis and performance tracking. Establishing rigorous data integration and preprocessing standards is crucial to mitigate these issues.

- **Data Accessibility**: Sometimes, the data needed to measure AI success is not readily accessible due to privacy concerns, regulatory restrictions, or technical barriers. Developing a framework that ensures data is accessible while complying with all legal and ethical guidelines is necessary for ongoing AI evaluation.

AI systems do not operate in a vacuum; they are deployed within ever-changing internal and external business environments that can significantly impact their performance and the relevance of the metrics used to measure their success.

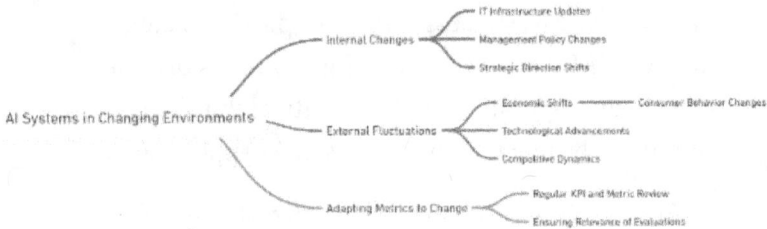

- Internal Changes: Organizational changes such as updates to IT infrastructure, changes in management policies, or shifts in strategic direction can alter the effectiveness of AI systems. For instance, a new IT policy might restrict the data AI systems can access, reducing their effectiveness or skewing performance metrics.

- External Fluctuations: Economic shifts, technological advancements, and competitive dynamics can also influence AI performance. An economic downturn might change consumer behavior patterns, affecting the predictions of AI models in marketing analytics. Similarly, new technologies may render existing AI tools obsolete or less effective.

- Adapting Metrics to Change: To maintain the reliability of AI measurements, it's necessary to regularly review and adjust KPIs and metrics to reflect current realities. This adaptive approach ensures that evaluations remain relevant and that AI systems continue to meet organizational needs even as conditions change.

Strategies for Overcoming Measurement Challenges

Implementing robust data governance practices is essential to address data complexity issues, ensuring high-quality, consistent, and accessible data for AI systems. Additionally, building flexibility into the AI measurement framework allows businesses to adapt metrics and KPIs in response to internal and external changes. Regular audits and updates to the measurement framework can help mitigate the impact of these dynamics on the evaluation of AI success.

By acknowledging and addressing these challenges, organizations can improve the accuracy and reliability of their AI success measurements, making these insights truly actionable for decision-making and strategic planning.

Demonstrating AI's Business Value

The primary value of implementing precise KPIs and metrics lies in their ability to demonstrate the concrete business benefits of AI technologies:

- **Quantifying Performance Improvements**: Metrics such as process efficiency, error reduction, and response times provide clear, quantitative data that showcases how AI improves operational workflows and outputs.

- **Validating Financial Returns**: Financial impact metrics, including ROI, cost savings, and revenue enhancements, directly link AI initiatives to the bottom line. These figures are crucial for justifying ongoing and future investments in AI technologies.

- **Enhancing Customer Experiences**: Customer-centric metrics help quantify how AI tools improve service delivery, customer satisfaction, and retention, which are vital for competitive differentiation.

Guiding Future AI Strategies

Beyond their role in validation, KPIs and metrics are invaluable for refining and guiding future AI strategies:

- **Strategic Alignment**: Metrics ensure that AI initiatives remain aligned with broader business objectives, providing a benchmark against which strategic adjustments can be made.

- **Identifying Opportunities for Improvement**: Continuous measurement helps identify underperforming areas and opportunities for enhancement, ensuring that AI systems evolve in response to operational needs and market conditions.

- **Facilitating Scalability**: By monitoring success at various scales, metrics can inform decisions on whether and how to expand AI applications across the organization.

Future Planning

The dynamic nature of AI technology and its applications means that the metrics used to measure it must also evolve. Keeping your KPIs and metrics up-to-date ensures that they remain relevant and continue to provide valuable insights. This adaptive measurement approach supports sustained innovation and growth, ensuring that AI continues to deliver strategic value in the long term.

AI OKR and KPI Checklist for Board and C-Suite

Scoring System

- **0 points:** Not addressed
- **1 point:** Partially addressed
- **2 points:** Fully addressed, but needs improvement
- **3 points:** Fully addressed and well-executed

General Considerations

1. **Alignment with Business Objectives** (Score: 0-3)
   - Are AI KPIs and OKRs aligned with the organization's strategic goals?

2. **Understanding of AI Project Objectives** (Score: 0-3)

    - Is there a clear understanding of the specific objectives of each AI project?

3. **Selection of Relevant Metrics** (Score: 0-3)

    - Are the selected KPIs relevant to the specific AI initiatives and their goals?

Efficiency Metrics

4. **Process Time Reduction** (Score: 0-3)

    - Are metrics in place to measure the reduction in time taken to complete tasks after AI implementation?

5. **Resource Utilization Rates** (Score: 0-3)

    - Are resource utilization rates being tracked to evaluate efficiency improvements?

6. **Reduction in Human Intervention** (Score: 0-3)

    - Is the decrease in human intervention for automated processes being measured?

Accuracy Metrics

7. **Correctness of Outputs** (Score: 0-3)

    - Are there accuracy metrics to evaluate the correctness of AI outputs?

8. **Error Rates** (Score: 0-3)

    - Are error rates being tracked before and after AI implementation to measure improvements?

Performance Metrics

9. **System Uptime** (Score: 0-3)

    - Is system uptime being measured to ensure reliability?

10. **Response Times** (Score: 0-3)

    - Are response times being tracked to evaluate the performance of AI systems?

11. **User Interaction Quality** (Score: 0-3)

    - Is the quality of user interactions with AI systems being assessed?

Financial Impact Metrics

12. **Return on Investment (ROI)** (Score: 0-3)

    - Is ROI being calculated to measure the financial benefits of AI initiatives?

13. **Cost Savings** (Score: 0-3)

    - Are cost savings from AI implementations being quantified?

14. **Revenue Contribution** (Score: 0-3)

    - Is the revenue generated from AI-enhanced products or services being tracked?

Customer Satisfaction Metrics

15. **Response Times in Customer Service** (Score: 0-3)
    - Are response times in customer service being measured to evaluate AI impact?

16. **Customer Service Quality** (Score: 0-3)
    - Is the quality of customer service being assessed through surveys and feedback?

17. **Customer Retention Rates** (Score: 0-3)
    - Are customer retention rates being tracked to measure the impact of AI on customer loyalty?

Strategic Implementation

18. **Clear Roles and Responsibilities** (Score: 0-3)
    - Are roles and responsibilities clearly defined for AI teams?

19. **Governance Framework** (Score: 0-3)
    - Is there a governance framework in place to ensure alignment with strategic goals and compliance with regulations?

20. **Training and Development** (Score: 0-3)
    - Are continuous training and development opportunities provided for AI teams?

Continuous Improvement

21. **Regular Monitoring and Audits** (Score: 0-3)
    - Are there regular audits and monitoring to ensure the effectiveness of AI KPIs and OKRs?

22. **Adapting Metrics to Change** (Score: 0-3)
    - Is there a system in place for regularly updating KPIs and metrics to reflect current realities?

Total Scoring

Each question is scored from 0 to 3, with a maximum possible score of 66 across all 22 questions.

Scoring and Interpretation

- **0–22**: High risk – Immediate action required to address significant gaps in AI KPI and OKR management.
- **23–44**: Moderate risk – Areas for improvement identified, and a structured plan is needed to enhance AI KPI and OKR management.
- **45–66**: Low risk – Robust AI KPI and OKR management practices in place, but continuous monitoring and minor adjustments are recommended.

Threshold for Passing

Organizations should aim for a minimum score of **45** to ensure they have adequately addressed key AI KPI and OKR considerations and are prepared to effectively measure and maximize the impact of their AI initiatives.

Summary

The integration and scaling of AI within an organization bring transformative potential, but the true measure of this technology's value lies in its tangible results. This is why the meticulous measurement of AI's impact through well-defined key performance indicators (KPIs) and metrics is indispensable. These tools not only quantify the success of AI initiatives but also guide strategic decision-making and future technology deployments. As we conclude, let's recap the critical role of these metrics and the value they bring to any organization investing in AI.

CHAPTER 14

AI Partnerships and Strategic Alliances

In the era of rapid technological advancements, artificial intelligence (AI) stands as a transformative force with the potential to revolutionize business landscapes. The economic impact of AI, particularly generative AI (gen AI), promises to be profound, driving productivity and innovation across sectors. However, capturing the full value of AI is not solely about the technology itself. Successful implementation and maximization of AI's potential require strategic partnerships and alliances. This chapter explores how companies can forge and manage these crucial relationships to harness AI's transformative power.

The Strategic Importance of AI Partnerships

AI's complexity and rapid evolution mean that companies cannot rely solely on internal capabilities to develop and deploy AI solutions effectively. Strategic partnerships with AI providers offer significant advantages, including access to cutting-edge technologies, specialized expertise, and accelerated development timelines. However, these alliances must be approached differently from traditional vendor relationships to truly unlock AI's potential.

CHAPTER 14   AI PARTNERSHIPS AND STRATEGIC ALLIANCES

Key Elements of Effective AI Partnerships

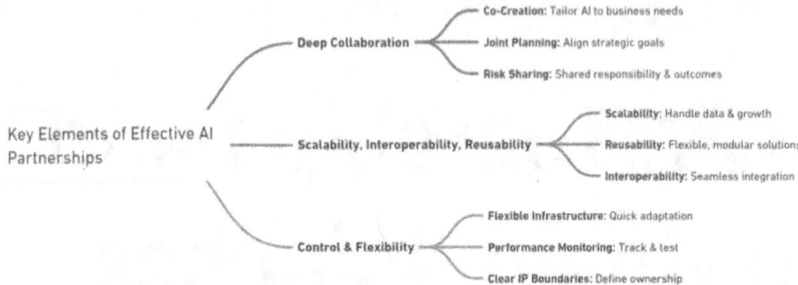

1. Deep Collaboration

Effective AI partnerships demand a higher degree of trust and collaboration than traditional vendor arrangements. Transparency, frequent communication, and alignment across planning, development, and ongoing management are crucial. Companies should engage in co-creation, joint planning, and risk and investment sharing to foster a collaborative environment.

- **Co-Creation of Solutions**: The most significant value from AI often comes from adopting established capabilities and tailoring them to unique business needs. This requires iterative collaboration, where both the company and the provider work closely to prepare data, engineer prompts, fine-tune models, and test solutions in real-world scenarios. Regular workshops, co-innovation sessions, and open sharing of insights help build the necessary trust and alignment.

- **Joint Planning**: Companies should have visibility into their AI providers' product roadmaps and engage in joint planning sessions. This ensures that both parties can anticipate future needs, influence development directions, and align their strategic goals. Collaborative planning helps integrate various AI models and applications, reducing time to market and enhancing solution effectiveness.

- **Risk and Investment Sharing**: AI initiatives often involve significant investments and risks. Companies should define clear responsibilities for mitigating and managing risks such as data privacy breaches, model biases, and intellectual property (IP) issues. Structured contracts that tie compensation to outcomes and align incentives can help manage uncertainties and foster shared commitment to success.

2. Scalability, Interoperability, and Reusability

Selecting AI providers that can scale and integrate seamlessly with existing systems is critical. Companies should focus on providers that offer scalable solutions, reusable components, and interoperable models.

- **Scalability**: Providers must demonstrate the ability to handle increasing volumes of data and user queries without compromising performance. Companies should evaluate scalability through pilot programs and ensure providers can meet specific milestones and adapt to evolving needs.

- **Reusability**: Reusable code and modular components can significantly accelerate AI development. Companies should seek providers offering flexible solutions that can be repurposed across various projects, enhancing development speed and reducing costs.

- **Interoperability**: Ensuring that AI models and components can work together seamlessly is vital. Providers should adhere to industry standards and best practices for data exchange and software development, facilitating smooth integration with the company's data sources and applications.

3. Maintaining Control and Flexibility

Balancing close strategic alliances with maintaining control over the direction and vision of AI initiatives is essential. Companies should establish flexible infrastructures, continually monitor performance, and set clear IP boundaries.

- **Flexible Infrastructure**: A scalable AI infrastructure, supported by machine-learning operations (MLOps) best practices, allows for quick integration of different providers and rapid adaptation to changes. Containerization, automated testing, and continuous integration and delivery (CI/CD) pipelines ensure reliability and performance.

- **Performance Monitoring**: Robust monitoring and testing capabilities are necessary to track provider performance and identify issues early. Regular end-to-end tests and comprehensive testing strategies involving all providers help maintain solution quality and transparency.

- **Clear IP Boundaries**: Defining IP ownership and management up front prevents disputes and ensures fair recognition of contributions. Companies should establish clear agreements on existing and co-developed IP, including licensing, commercialization, and revenue sharing terms.

CHAPTER 14   AI PARTNERSHIPS AND STRATEGIC ALLIANCES

Getting Started with AI Partnerships

To successfully navigate AI partnerships and strategic alliances, companies should consider the following actions:

- **Establish a Steering Committee**: Form a committee of key stakeholders from business, IT, legal, and procurement to oversee the AI alliance strategy. This committee should define strategic criteria, set performance metrics, and establish governance guidelines.

- **Develop a Strategic Alliance Playbook**: Create a standardized framework for evaluating, onboarding, and managing AI providers. This playbook should include guidelines for assessing scalability, reusability, and interoperability, along with templates for contracts and performance dashboards.

- **Conduct a Strategic Alliance Audit**: Assess current alliances to identify gaps, redundancies, or misalignments with the AI strategy. Determine which alliances to maintain, expand, or phase out based on their potential to drive business value.

- **Assign Relationship Managers**: Designate dedicated managers with a solid understanding of AI technologies to oversee provider relationships, ensure alignment with technical requirements, and coordinate activities among providers.

Summary

Building strategic AI partnerships is a critical component of leveraging AI's full potential. These alliances enable companies to access advanced technologies, specialized expertise, and scalable solutions, driving innovation and competitive advantage. By fostering deep collaboration, ensuring scalability and interoperability, and maintaining control and flexibility, companies can unlock significant value from their AI initiatives.

As the AI landscape continues to evolve, companies that strategically position themselves through effective partnerships will be better equipped to navigate the complexities of AI transformation and capitalize on the opportunities it presents. By acting decisively and building robust, collaborative relationships, organizations can ensure they are at the forefront of the AI revolution, driving sustained growth and innovation.

CHAPTER 15

AI Talent Strategy

Artificial intelligence (AI) is not just another technological advancement; it is a paradigm shift that is reshaping industries and redefining competitive landscapes. The promise of AI is vast, but realizing its potential requires more than just the latest technology. It necessitates a skilled workforce capable of driving innovation and integrating AI into business processes.

However, the scarcity of AI talent poses a significant challenge. To overcome this, organizations must develop a robust AI talent strategy that encompasses attracting, developing, and retaining top talent.

The AI Talent Landscape

The demand for AI professionals has skyrocketed, yet the supply remains critically low. This imbalance has led to fierce competition among organizations to secure the best talent. To navigate this competitive environment, companies must adopt a comprehensive strategy that goes beyond traditional hiring practices.

CHAPTER 15   AI TALENT STRATEGY

Attracting AI Talent

To attract AI talent, organizations need to understand what these professionals value most in their careers. Unlike traditional job seekers, AI candidates prioritize opportunities to work on cutting-edge projects and seek clear career advancement paths.

1. Define a Clear Value Proposition

Articulate the unique opportunities and career paths available within your organization. Highlight involvement in innovative projects and the potential for professional growth. This not only attracts talent but also sets the stage for their long-term engagement.

2. Leverage Untapped Talent Pools

Expand your search beyond traditional tech hubs. While major cities like San Francisco and New York are saturated, secondary markets and remote work options can provide access to exceptional talent. Many AI professionals are open to remote work, allowing you to tap into a global talent pool without geographic constraints.

3. Customize Recruiting Processes

Traditional recruiting processes are often too slow and generic for AI talent. Streamline your hiring processes and involve AI experts in interviews. Quick follow-ups and a smooth recruitment process can secure top candidates before they are snapped up by competitors.

4. Anchor Hires

For organizations new to AI, starting with an anchor hire – a highly skilled AI professional who can attract a network of specialists – can establish a strong foundation. This approach builds credibility and attracts additional talent.

Developing AI Talent

With the scarcity of external talent, developing AI capabilities internally is both a strategic necessity and a competitive advantage. Internal development involves reskilling and upskilling current employees, creating a pipeline of AI-ready talent within the organization.

CHAPTER 15   AI TALENT STRATEGY

1. Reskilling Programs

Invest in reskilling programs to transition current employees into AI roles. These workers already understand your business and can quickly adapt to new technical skills, providing a valuable bridge between existing operations and new AI initiatives.

2. Continuous Learning Culture

Establish a culture of continuous learning. Regular training, certifications, and learning opportunities ensure employees stay updated with the latest advancements and techniques. This commitment to learning fosters innovation and keeps your workforce agile.

3. Structured Career Paths

Define clear career paths for AI professionals. Frequent promotions and opportunities to tackle complex projects keep AI employees engaged and motivated. A well-structured career progression plan can be a significant retention tool.

4. Communities of Practice

Create internal communities where AI professionals can share knowledge, collaborate on projects, and support each other. These communities foster a sense of belonging and professional development, spreading best practices across the organization.

Retaining AI Talent

Retention of AI talent requires more than competitive salaries; it involves creating an environment where employees feel valued, challenged, and connected to the organization's mission.

1. Purpose-Driven Work

Communicate how AI projects align with broader company goals and societal benefits. AI professionals are motivated by the impact of their work. Reinforce the importance of their contributions to maintain high levels of engagement.

2. Integration into the Organization

Avoid isolating AI teams. Integrate them into cross-functional teams and involve them in strategic decision-making processes. This ensures their work is aligned with business objectives and enhances their sense of purpose.

3. Flexible Work Arrangements

Offer flexible work arrangements, including remote work and flexible hours. This flexibility is particularly important for attracting and retaining top talent who prioritize work–life balance.

4. Recognition and Rewards

Regularly recognize and reward contributions. This can be in the form of financial incentives or public acknowledgment of achievements. Consistent recognition boosts morale and motivation.

CHAPTER 15 AI TALENT STRATEGY

Strategic Considerations for AI Talent Management

Successfully managing AI talent requires a strategic approach that encompasses the entire employee lifecycle – from attraction to development to retention. Key considerations include

1. Data-Driven Decisions

Leverage data to understand talent needs, track performance, and identify gaps. Digital employee profiles and analytics provide insights into current skills and future requirements, ensuring your talent strategy remains aligned with organizational goals.

2. Collaborative Ecosystem

Build a collaborative ecosystem with academic institutions, industry partners, and AI communities. These partnerships provide access to a broader talent pool and foster innovation. Collaborating with educational organizations can also help shape curricula to better prepare future AI professionals.

3. Ethical and Responsible AI

Ensure that AI is developed and deployed ethically. Prioritize diversity in AI teams to bring multiple perspectives and mitigate biases. Establish clear guidelines for responsible AI use to build trust with employees and stakeholders.

4. Long-Term Vision

Anticipate how AI will impact the organization and plan accordingly. This includes evolving the organizational structure, redesigning processes, and preparing the workforce for continuous technological advancements.

Case Study: Booz Allen Hamilton

Booz Allen Hamilton provides a compelling example of a successful AI talent strategy. The firm has built a robust AI workforce by focusing on several key areas:

1. Early Adoption and Centralized Teams

Booz Allen identified AI as a strategic priority early on, centralizing a team around data science in the early 2010s. This team has grown to support various sectors, from healthcare to defense.

2. Proactive Talent Mapping

Using tools like digital employee profiles, Booz Allen maps current talent against future needs, ensuring they are prepared for evolving demands.

3. Partnerships with Educational Institutions

Booz Allen partners with organizations such as aiEDU and The Mark Cuban Foundation to incorporate AI into curriculums and prepare future talent from a young age.

4. Comprehensive Training Programs

Programs for reskilling and upskilling are integral to Booz Allen's strategy. Their Tech Excellence Program and digital badging initiatives ensure that employees at all levels have access to AI training.

Moving Forward: Implementing an AI Talent Strategy

As AI continues to evolve, organizations must act decisively to position themselves for success. Here are actionable steps to implement an effective AI talent strategy:

1. Establish a Steering Committee

Form a committee of key stakeholders from business, IT, legal, and procurement to oversee the AI talent strategy. This committee should define strategic criteria, set performance metrics, and establish governance guidelines.

2. Develop a Strategic Talent Playbook

Create a standardized framework for evaluating, onboarding, and managing AI talent. This playbook should include guidelines for assessing skills, creating structured career paths, and implementing training programs.

3. Conduct a Talent Audit

Assess current talent to identify gaps, redundancies, and alignment with the AI strategy. Determine which areas need new hires, reskilling, or structural changes to support AI initiatives.

4. Assign Dedicated Relationship Managers

Designate managers with a deep understanding of AI to oversee talent development and integration. These managers will ensure alignment with technical requirements and coordinate activities among different teams.

5. Foster a Culture of Innovation

Encourage experimentation and innovation within your AI teams. Provide resources and support for employees to explore new ideas and approaches. This fosters a dynamic environment where talent can thrive.

6. Emphasize Ethical AI

Ensure all AI initiatives adhere to ethical standards. Promote transparency, fairness, and accountability in AI development and deployment. This not only builds trust but also attracts professionals who value ethical practices.

7. Monitor and Adapt

Continuously monitor the effectiveness of your AI talent strategy and be prepared to adapt. Regularly review metrics, gather feedback, and make necessary adjustments to stay aligned with evolving business needs and technological advancements.

Generative AI: Redefining Leadership

As generative AI (Generative AI) becomes increasingly integral to business operations, the roles of C-suite executives are evolving. This chapter outlines the specific responsibilities and competencies required for key leadership positions to effectively harness the potential of Generative AI while managing associated risks.

CHAPTER 15 AI TALENT STRATEGY

Chief Executive Officer (CEO)
Responsibilities

The CEO bears the ultimate responsibility for ensuring that Generative AI initiatives align with the company's strategic goals and maintain high standards of ethical practice and customer trust. This involves leading the development of clear objectives for Generative AI across the organization and ensuring these initiatives deliver tangible business value.

Competencies and Actions

- **Understanding of Generative AI:** CEOs must possess a foundational knowledge of Generative AI, focusing on security and privacy risks to hold other C-suite members accountable.

- **Strategic Leadership:** Develop and enforce a formal Generative AI operating model that includes protocols for engaging with third-party vendors and platform providers.

- **Collaborative Oversight:** Ensure that product, legal, and security teams collaboratively manage all phases of Generative AI deployment. CEOs must also manage internal conflicts effectively and maintain alignment across business units.

Chief Operating Officer (COO)

Responsibilities

The COO ensures that Generative AI deployments are robust and align with the organization's operational goals. This role is critical in embedding Generative AI technologies into daily operations without disrupting the existing workflows.

Competencies and Actions

- **Operational Alignment:** Establish Generative AI objectives that integrate with business operations and include measurable security and privacy processes.

- **Resilience Management:** Develop capabilities to support Generative AI operations and maintain resilience against potential disruptions.

- **Troubleshooting:** Maintain oversight of operational challenges and ensure the organization can address and resolve disruptions in Generative AI-enabled processes swiftly.

Responsibilities of CROs and CISOs

- **Risk Identification and Strategy Formulation:** They are tasked with identifying potential threats introduced by Generative AI technologies and developing strategic measures to mitigate these risks.

- **Governance and Compliance:** Ensuring that Generative AI implementations adhere to established data governance standards and comply with relevant laws and regulations.

- **Vendor Management:** Managing relationships with Generative AI vendors involves rigorous vetting processes to ensure that their practices align with the company's security standards.

Updated Competencies and Actions

- **Developing a Generative AI-Specific Security Strategy:** They must establish a security strategy that defines clear risk tolerance levels for Generative AI initiatives.

- **Operational Collaboration**: Working in tandem with the COO to set measurable criteria for assessing Generative AI use cases against their potential value and risks.

- **Skill Development and Talent Management**: There is an urgent need to reskill or upskill security personnel to prepare them for the unique challenges posed by Generative AI, ensuring the team's capabilities evolve as rapidly as the technologies they oversee.

Additional Strategic Focus Areas

- **Security by Design**: Updating and implementing secure-by-design standards throughout the machine-learning operations process, ensuring that all Generative AI applications are built with security at the forefront

- **Incident Response Preparedness**: Establishing and refining incident response protocols to address potential security breaches or regulatory issues swiftly and effectively

- **Enhanced Cybersecurity Measures with Generative AI**: Evaluating and utilizing Generative AI-driven cybersecurity solutions to enhance the organization's ability to detect and respond to threats

Chief Information, Technology, and Data Officers (CIOs, CTOs, CDOs)

Responsibilities

- Ensure the organization has access to the necessary technology infrastructure, systems, applications, services, and data to support Generative AI initiatives effectively.

- Maintain a delicate balance between cost and value, ensuring Generative AI implementations enhance rather than hinder innovation.

Competencies and Actions

- **Data Management:** Monitor data provenance and security, ensuring Generative AI systems manage and utilize data without compromising security classifications.

- **Development and Experimentation:** Establish environments conducive to Generative AI development, guiding the creation and use of Generative AI technologies within the organization.

- **Interdepartmental Collaboration:** Work closely with risk, legal, and security teams to align Generative AI initiatives with organizational policies and risk management strategies.

Chief Legal and Privacy Officers

Responsibilities

- Ensure Generative AI adoption complies with legal standards and privacy regulations.
- Oversee data governance and manage Generative AI processes to align with legal requirements without stifling innovation.

Competencies and Actions

- **Regulatory Compliance:** Stay updated on new legislation affecting Generative AI and implement practical standards and procedures.
- **Monitoring and Reporting:** Establish and maintain mechanisms to ensure compliance with legal standards and internal procedures.

Chief Product Officer

Responsibilities

- Ensure that Generative AI applications add substantive value to products, services, or processes, aligning with new customer value propositions or cost reductions.

Competencies and Actions

- **Innovation Management:** Collaborate with risk and security officers to integrate responsible AI and secure-by-design practices within product teams.

- **Performance Metrics:** Develop and monitor key performance indicators to evaluate the impact and value of Generative AI-enabled products.

Chief Marketing Officer

Responsibilities

Oversee the creation of marketing materials using Generative AI, ensuring all content is legally compliant and does not infringe on intellectual property rights.

Competencies and Actions

- **Content Integrity:** Manage the risks associated with Generative AI-generated content, ensuring all materials are free from unauthorized content.

Chief Financial Officer

Responsibilities

- Manage the financial aspects of Generative AI projects, including budgeting, cost management, and evaluation of technology investments.

Competencies and Actions

- **Financial Oversight:** Understand and manage the costs associated with Generative AI, negotiating contracts that include terms for cybersecurity and privacy.
- **Investment Analysis:** Require comprehensive business cases for Generative AI investments that integrate security and privacy considerations.

Chief Human Resources Officer

Responsibilities

Develop and enforce training and personnel policies related to Generative AI to ensure all employees are equipped to utilize these technologies effectively.

Competencies and Actions

Training Programs: Coordinate with IT and security departments to provide necessary training on Generative AI tools and policies.

AI Talent Strategy Checklist for Board and C-Suite

Scoring System

- **0 points:** Not addressed
- **1 point:** Partially addressed

- **2 points:** Fully addressed, but needs improvement
- **3 points:** Fully addressed and well-executed

Attracting AI Talent

1. **Define a Clear Value Proposition** (Score: 0-3)
   - Have we articulated unique opportunities and career paths within the organization?
   - Are innovative projects and professional growth potential highlighted?

2. **Leverage Untapped Talent Pools** (Score: 0-3)
   - Do we expand our search beyond traditional tech hubs?
   - Are remote work options available to tap into a global talent pool?

3. **Customize Recruiting Processes** (Score: 0-3)
   - Is our hiring process streamlined and quick?
   - Are AI experts involved in interviews and follow-ups?

4. **Anchor Hires** (Score: 0-3)
   - Have we made strategic anchor hires to attract additional talent?
   - Does this approach build credibility within our AI team?

CHAPTER 15 AI TALENT STRATEGY

Developing AI Talent

5. **Reskilling Programs** (Score: 0-3)
   - Are we investing in reskilling programs to transition current employees into AI roles?
   - Do these employees have a clear understanding of both business and technical skills?

6. **Continuous Learning Culture** (Score: 0-3)
   - Do we establish a culture of continuous learning?
   - Are regular training, certifications, and learning opportunities provided?

7. **Structured Career Paths** (Score: 0-3)
   - Are there clear career paths defined for AI professionals?
   - Are frequent promotions and complex project opportunities available?

8. **Communities of Practice** (Score: 0-3)
   - Do we have internal communities for AI professionals to share knowledge and collaborate?
   - Do these communities foster a sense of belonging and professional development?

Retaining AI Talent

9. **Purpose-Driven Work** (Score: 0-3)
   - Do we communicate how AI projects align with broader company goals and societal benefits?

- Are AI professionals motivated by the impact of their work?

10. **Integration into the Organization** (Score: 0-3)

    - Are AI teams integrated into cross-functional teams?
    - Are they involved in strategic decision-making processes?

11. **Flexible Work Arrangements** (Score: 0-3)

    - Do we offer flexible work arrangements, including remote work and flexible hours?
    - Does this flexibility cater to top talent's need for work-life balance?

12. **Recognition and Rewards** (Score: 0-3)

    - Are contributions regularly recognized and rewarded?
    - Are financial incentives and public acknowledgment of achievements provided?

Strategic Considerations for AI Talent Management

13. **Data-Driven Decisions** (Score: 0-3)

    - Do we leverage data to understand talent needs, track performance, and identify gaps?
    - Are digital employee profiles and analytics used for insights into skills and requirements?

14. **Collaborative Ecosystem** (Score: 0-3)
    - Do we build a collaborative ecosystem with academic institutions, industry partners, and AI communities?
    - Do these partnerships provide access to a broader talent pool and foster innovation?

15. **Ethical and Responsible AI** (Score: 0-3)
    - Do we prioritize diversity in AI teams to bring multiple perspectives and mitigate biases?
    - Are there clear guidelines for responsible AI use to build trust with employees and stakeholders?

16. **Long-Term Vision** (Score: 0-3)
    - Do we anticipate how AI will impact the organization and plan accordingly?
    - Are organizational structures, processes, and workforce prepared for continuous technological advancements?

Implementation and Monitoring

17. **Establish a Steering Committee** (Score: 0-3)
    - Is there a committee of key stakeholders to oversee the AI talent strategy?
    - Does this committee define strategic criteria, set performance metrics, and establish governance guidelines?

CHAPTER 15 AI TALENT STRATEGY

18. **Develop a Strategic Talent Playbook** (Score: 0-3)

    - Do we have a standardized framework for evaluating, onboarding, and managing AI talent?

    - Are there guidelines for assessing skills, creating structured career paths, and implementing training programs?

19. **Conduct a Talent Audit** (Score: 0-3)

    - Have we assessed current talent to identify gaps, redundancies, and alignment with the AI strategy?

    - Are areas needing new hires, reskilling, or structural changes identified?

20. **Assign Dedicated Relationship Managers** (Score: 0-3)

    - Are managers with a deep understanding of AI overseeing talent development and integration?

    - Do these managers ensure alignment with technical requirements and coordinate activities among different teams?

21. **Foster a Culture of Innovation** (Score: 0-3)

    - Do we encourage experimentation and innovation within our AI teams?

    - Are resources and support provided for employees to explore new ideas and approaches?

22. **Emphasize Ethical AI** (Score: 0-3)

    - Are all AI initiatives adhering to ethical standards?

    - Is transparency, fairness, and accountability promoted in AI development and deployment?

23. **Monitor and Adapt** (Score: 0-3)

    - Are we continuously monitoring the effectiveness of our AI talent strategy?
    - Are there regular reviews of metrics, feedback gathering, and necessary adjustments made?

Total Scoring

Each question is scored from 0 to 3, with a maximum possible score of 69 across all 23 questions.

Scoring and Interpretation

- **0–23**: High risk – Immediate action required to address significant gaps in AI talent strategy.
- **24–46**: Moderate risk – Areas for improvement identified, and a structured plan is needed to enhance AI talent management.
- **47–69**: Low risk – Robust AI talent management practices in place, but continuous monitoring and minor adjustments are recommended.

Threshold for Passing

Organizations should aim for a minimum score of **47** to ensure they have adequately addressed key AI talent strategy considerations and are prepared to attract, develop, and retain top AI talent effectively.

Summary

Crafting a robust AI talent strategy is essential for any organization aiming to harness the full potential of artificial intelligence. By attracting, developing, and retaining top AI talent, companies can ensure they are well-positioned to lead in this transformative era. A comprehensive and forward-thinking approach, integrating AI talent into the organization's fabric and fostering an environment of continuous learning, innovation, and ethical practices, will be the cornerstone of sustained competitive advantage and long-term success. As the AI landscape continues to evolve, a robust talent strategy will not only drive business transformation but also pave the way for a future where AI is seamlessly integrated into every aspect of the enterprise.

CHAPTER 16

AI Monetization: Strategies for Profitable Innovation

Artificial Intelligence (AI) is not just a technological advancement; it represents a seismic shift in how businesses operate and compete. The challenge lies in monetizing AI effectively to ensure it generates substantial revenue while aligning with broader business goals. This chapter delves into the various strategies and business models that organizations can adopt to monetize AI, leveraging its full potential to drive growth and profitability.

AI-as-a-Service (AIaaS)

AI-as-a-Service (AIaaS) is revolutionizing how businesses access and utilize AI capabilities. This model parallels the success of Software-as-a-Service (SaaS) by providing scalable and flexible AI solutions via the cloud.

1. Subscription Models

Subscription-based AIaaS is a popular approach, offering predictable revenue streams and affordability for users. Companies like OpenAI provide monthly subscriptions for access to their AI tools, allowing continuous and seamless use of AI technologies.

2. Pay-Per-Use Models

The pay-per-use model charges users based on their actual usage, providing cost efficiency and flexibility. OpenAI, for instance, charges by "tokensv" used in data processing, making this model ideal for businesses with varying AI needs.

3. Freemium to Premium

Freemium models attract users by offering basic AI services for free, with premium features available for a fee. This approach allows users to experience the value of AI before committing to a paid plan, increasing user engagement and conversion rates.

Custom AI Solutions

Tailoring AI solutions to specific industry needs can significantly enhance their effectiveness and value, leading to greater monetization opportunities.

1. Enterprise Customization

Customizing AI models with an enterprise's proprietary data can yield highly relevant and effective solutions. This approach is particularly useful in sectors like healthcare, finance, and customer service, where tailored AI can drive significant improvements in performance and efficiency.

2. Confidentiality and Data Security

While customization offers immense benefits, it also raises concerns about data confidentiality and security. These concerns can be managed through robust data governance frameworks, ensuring secure and compliant handling of sensitive information.

3. Third-Party Service Providers

Specialized service providers that curate datasets and train AI models for specific corporate applications play a crucial role. These providers can offer bespoke solutions that meet unique business needs, opening new revenue streams.

AI-Powered Products and Services

The integration of AI into products and services presents substantial monetization opportunities, enhancing value propositions and driving competitive advantage.

1. Enhancing Existing Products

Incorporating AI into existing products can enhance their functionality and user experience. For example, AI-driven features in software products can automate complex tasks, provide deeper insights, and improve overall user satisfaction.

2. Developing New Offerings

AI enables the creation of new products and services that were previously unattainable. Startups, in particular, can leverage AI to innovate and compete with larger firms by offering cutting-edge solutions.

3. Market Differentiation

AI can serve as a key differentiator in crowded markets. Companies that effectively integrate AI into their offerings can stand out by providing superior performance, personalization, and customer engagement.

Legal and Ethical Considerations

Monetizing AI involves navigating complex legal and ethical landscapes. Addressing these considerations is essential to ensure compliance and build trust.

1. Data Privacy and Security

Collecting and using data for AI requires stringent compliance with data protection regulations like GDPR and CCPA. Implementing robust data security measures is crucial to protect sensitive information from breaches.

2. Intellectual Property (IP)

Clear agreements on IP rights are vital, especially when working with third-party AI service providers. These agreements should specify ownership of data, models, and resulting insights to prevent disputes.

3. Liability and Indemnification

Defining liability and indemnification in service agreements is essential due to the uncertainties in AI performance. These agreements should outline responsibilities, performance guarantees, and the scope of liability.

Insurance as a Risk Management Tool

Commercial insurance can mitigate the risks associated with AI monetization. Customized insurance policies can cover potential liabilities, providing a safety net for both AI providers and their clients.

1. Bridging Liability Gaps

Insurance can help bridge the gap between customer expectations and the limitations of AI service providers. This transfer of financial risk to insurers allows for more favorable contractual terms.

2. Customized Policies

Insurers can develop tailored policies that address specific AI-related risks, such as data breaches and algorithm errors. These policies ensure comprehensive coverage and enhance trust in AI solutions.

Future Trends in AI Monetization

The landscape of AI monetization is continuously evolving. Emerging trends and technologies will shape how businesses capitalize on AI in the future.

1. Open Source AI

Open-source AI projects offer new monetization opportunities by reducing development costs and accelerating innovation. Companies can leverage these solutions to enhance their offerings and reduce time-to-market.

2. Data As a Competitive Advantage

As core AI technologies become commoditized, the quality and exclusivity of training data will be key differentiators. Companies with access to large, high-quality datasets will have a significant edge.

3. AI Ecosystems

Building comprehensive AI ecosystems that integrate various AI capabilities will become increasingly important. These ecosystems provide end-to-end solutions, enhancing value for customers and creating new revenue opportunities.

Case Study: OpenAI

OpenAI provides a compelling example of effective AI monetization through its dual-approach model for ChatGPT. By offering both a monthly subscription and a pay-per-use API license, OpenAI caters to diverse customer needs, ensuring accessibility and scalability.

1. Subscription and Pay-Per-Use Models

The subscription model offers predictable revenue, while the pay-per-use model aligns costs with actual usage, providing flexibility for businesses with varying needs.

2. Token-Based Pricing

OpenAI's use of tokens for pay-per-use pricing allows precise cost management, making it easier for businesses to budget for AI expenses.

3. Freemium Model

The freemium model attracts users with free basic services, encouraging them to upgrade to premium plans as they experience the value of the AI tools.

AI Monetization Consideration Checklist for Board and C-Suite

To assist in refining the selection and implementation of AI monetization strategies, this checklist focuses on common themes essential for any monetization approach. Subsequently, specific questions related to different AI monetization models are provided, acknowledging that an organization may adopt one or two models based on their strategic objectives.

Scoring System

- **0 points:** Not addressed
- **1 point:** Partially addressed
- **2 points:** Fully addressed, but needs improvement
- **3 points:** Fully addressed and well-executed

CHAPTER 16   AI MONETIZATION: STRATEGIES FOR PROFITABLE INNOVATION

Common Themes

1. **Strategic Alignment** (Score: 0-3 each)
   - Does the monetization strategy align with the organization's broader strategic goals?
   - Are the AI initiatives integrated into the overall business strategy to ensure coherence and support?

2. **Governance and Compliance** (Score: 0-3 each)
   - Are there robust governance frameworks in place to oversee AI monetization efforts?
   - Do we comply with relevant legal and ethical standards, such as data privacy regulations?

3. **Data Management** (Score: 0-3 each)
   - Is there a strong data management framework to ensure data quality, security, and privacy?
   - Are data governance policies in place to handle proprietary and third-party data?

4. **Risk Management** (Score: 0-3 each)
   - Have potential risks associated with AI monetization been identified and mitigated?
   - Are insurance policies or other risk management tools in place to cover AI-related liabilities?

5. **Performance Monitoring** (Score: 0-3 each)
   - Are there metrics and KPIs to monitor the performance and ROI of AI initiatives?
   - Is there a system for continuous improvement and optimization of AI processes?

CHAPTER 16   AI MONETIZATION: STRATEGIES FOR PROFITABLE INNOVATION

Monetization Models
AI-as-a-Service (AIaaS)

6. **Subscription Models** (Score: 0-3 each)
   - Are we offering AI services through a subscription model for predictable revenue?
   - Do we provide affordable subscription plans that attract a broad user base?

7. **Pay-Per-Use Models** (Score: 0-3 each)
   - Do we offer a pay-per-use model that charges based on actual usage?
   - Are cost efficiency and flexibility maintained for varying business needs?

8. **Freemium to Premium** (Score: 0-3 each)
   - Do we use a freemium model to attract users with basic AI services for free?
   - Are premium features available for a fee to encourage upgrades and higher engagement?

Custom AI Solutions

9. **Enterprise Customization** (Score: 0-3 each)
   - Are AI models customized with proprietary data for specific industries?
   - Do these customizations drive significant improvements in performance and efficiency?

10. **Third-Party Service Providers** (Score: 0-3 each)
    - Do we collaborate with specialized providers to curate datasets and train AI models?
    - Are bespoke solutions developed to meet unique business needs and open new revenue streams?

AI-Powered Products and ServicesEnhancing Existing Products (Score: 0-3 each)

- Is AI integrated into existing products to enhance functionality and user experience?
- Are AI-driven features automating complex tasks and improving overall satisfaction?

11. **Developing New Offerings** (Score: 0-3 each)
    - Are new AI-enabled products and services being developed to tap into untapped markets?
    - Do these innovations offer cutting-edge solutions that provide competitive advantages?

12. **Market Differentiation** (Score: 0-3 each)
    - Is AI used as a differentiator in crowded markets to offer superior performance?
    - Are AI-driven personalization and customer engagement strategies enhancing our market position?

Total Scoring

Each question is scored from 0 to 3, with a maximum possible score of 48 across all 16 questions (common themes and two monetization models).

Scoring and Interpretation

- **0-16**: High risk – Immediate action required to address significant gaps in AI monetization strategies.

- **17-32**: Moderate risk – Areas for improvement identified, and a structured plan is needed to enhance AI monetization approaches.

- **33-48**: Low risk – Robust AI monetization strategies in place, but continuous monitoring and minor adjustments are recommended.

Threshold for Passing

Organizations should aim for a minimum score of **32** to ensure they have adequately addressed key AI monetization considerations and are prepared to manage and scale AI operations effectively

Summary

Monetizing AI requires a strategic approach that combines innovative business models with robust legal and ethical frameworks. By exploring various monetization strategies, customizing AI solutions, and staying ahead of emerging trends, companies can unlock the full potential of AI. Embracing AI as a core component of business strategy will drive growth, enhance competitive advantage, and ensure long-term success in the digital age.

CHAPTER 17

Aligning AI Investments with Business Problems

In crafting an effective AI strategy, it's critical to align investments with the scope and complexity of business problems. This chapter delves into a strategic model that categorizes potential AI applications based on their scope and the required level of investment, ensuring that resources are optimized and business impacts are maximized.

Understanding the Problem–Investment Matrix

The problem–investment matrix is a visual tool that helps categorize AI initiatives into tiers based on the scope of the problem they address and the investment they require. This matrix serves as a roadmap for businesses to strategically deploy AI solutions where they are most needed and can deliver the most value.

CHAPTER 17   ALIGNING AI INVESTMENTS WITH BUSINESS PROBLEMS

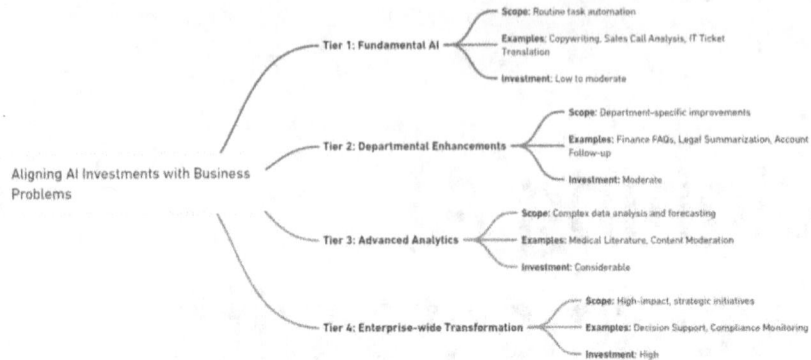

Tier 1: Fundamental AI Applications

Scope: These are AI applications that address basic, yet significant business processes. They typically involve automation of routine tasks.

Examples:

Copywriting for Websites: Automating the generation of content for websites, which can enhance speed and maintain brand consistency.

Sales Call Sentiment Analysis: Utilizing AI to analyze the tone and sentiment of sales calls to refine sales strategies and improve customer relations.

Ticket Translation for IT Agents: Facilitating quicker resolution of IT issues by translating tickets submitted in various languages.

Investment Level: Low to moderate. These solutions require relatively lower financial outlay but bring significant efficiency improvements.

Tier 2: Departmental Enhancements

Scope: These applications target specific departmental needs, improving accuracy and efficiency with a moderate level of complexity.

Examples:

Finance Department FAQs: Deploying AI to automatically answer frequently asked questions in finance, reducing workload and response times.

Legal Department Document Summarization: Utilizing AI to summarize lengthy legal documents, aiding in faster decision-making and increased productivity.

Account Team Follow-up: Automating follow-up tasks for account management teams to ensure timely interactions with clients.

Investment Level: Moderate. These solutions often require more customized AI tools and integration with existing departmental workflows.

Tier 3: Advanced Analytical Tools

Scope: Addressing more complex problems that involve significant amounts of data interpretation or operational forecasting.

Examples:

Medical Literature Analysis: AI-driven analysis of vast amounts of medical texts to support research and development.

Automated Content Moderation: Employing AI to monitor and moderate content across platforms, ensuring compliance with standards and reducing manual oversight.

Investment Level: Considerable. These applications need advanced AI capabilities and integration with complex systems.

Tier 4: Enterprise-wide Transformation

Scope: These are strategic, high-impact AI initiatives that affect the entire enterprise and require substantial investment.

Examples:

Decision-Making Support: Integrating AI into the decision-making processes to provide insights derived from comprehensive data analysis.

Compliance and Security Monitoring: Utilizing AI for ongoing monitoring of compliance and security, crucial for maintaining corporate integrity and avoiding legal issues.

Investment Level: High. These initiatives are strategic investments that often involve substantial resource allocation, including high financial stakes and significant changes to organizational processes.

Checklist for Aligning AI Investments with Business Problems for Board and C-Suite

This checklist is designed to help boards and C-suite executives ensure that AI investments are strategically aligned with the complexity and scope of business problems. The questions are quantitative, with each question being scored individually. A scoring system is provided to help determine the effectiveness of AI investment alignment.

Scoring System

- **0 points:** Not addressed
- **1 point:** Partially addressed
- **2 points:** Fully addressed, but needs improvement
- **3 points:** Fully addressed and well-executed

General Considerations

1. **Strategic Alignment** (Score: 0-3 each)
    - Is the AI investment aligned with the organization's overall strategic goals?
    - Are the selected AI initiatives focused on solving core business problems?

2. **Problem Scope Clarity** (Score: 0-3 each)

   - Is the scope of the business problem clearly defined and understood?
   - Does the scope align with the potential impact of the AI solution?

3. **Investment Justification** (Score: 0-3 each)

   - Is there a clear business case for the AI investment, including ROI and value proposition?
   - Are the financial and resource investments proportionate to the expected business benefits?

Tier 1: Fundamental AI Applications

4. **Automation Potential** (Score: 0-3 each)

   - Does the AI application target routine tasks that can be effectively automated?
   - Is the investment level appropriate for the simplicity and scope of the problem?

5. **Efficiency Improvement** (Score: 0-3 each)

   - Will the AI solution significantly improve operational efficiency for basic processes?
   - Are there measurable efficiency gains expected from the implementation?

Tier 2: Departmental Enhancements

6. **Departmental Impact** (Score: 0-3 each)
   - Is the AI initiative designed to address specific departmental challenges?
   - Does the solution integrate well with existing departmental workflows?

7. **Customization and Integration** (Score: 0-3 each)
   - Is there a need for customized AI tools tailored to departmental needs?
   - Are the integration efforts with existing systems adequately planned and resourced?

8. **Productivity Enhancement** (Score: 0-3 each)
   - Will the AI solution enhance productivity within the department?
   - Are there clear KPIs set to measure productivity improvements?

Tier 3: Advanced Analytical Tools

9. **Complex Problem Solving** (Score: 0-3 each)
   - Does the AI application address complex problems requiring advanced data interpretation or forecasting?
   - Is the investment level justified by the complexity and potential impact of the AI solution?

10. **Data Utilization** (Score: 0-3 each)

    - Is the AI solution capable of processing and analyzing large datasets to generate actionable insights?

    - Are data sources reliable, and is data governance in place to ensure quality?

11. **Innovation and R&D Support** (Score: 0-3 each)

    - Does the AI solution contribute to research and development or other innovation-driven initiatives?

    - Is there a clear link between AI-driven insights and strategic business decisions?

Tier 4: Enterprise-wide Transformation

12. **Strategic Impact** (Score: 0-3 each)

    - Is the AI initiative expected to drive significant changes across the entire enterprise?

    - Are there clear, strategic outcomes identified that align with enterprise-wide goals?

13. **Resource Allocation** (Score: 0-3 each)

    - Are resources, including financial, technological, and human, adequately allocated for this high-impact AI initiative?

    - Is there executive-level commitment to support the initiative through to completion?

14. **Change Management** (Score: 0-3 each)
    - Is there a robust change management strategy in place to handle organizational adjustments due to the AI implementation?
    - Are there clear communication plans to ensure alignment across all levels of the organization?

15. **Compliance and Security** (Score: 0-3 each)
    - Does the AI solution include ongoing monitoring for compliance and security to protect corporate integrity?
    - Are there safeguards in place to prevent legal and ethical issues?

Total Scoring

Each question is scored from 0 to 3, with a maximum possible score of 45 across all 15 questions.

Scoring and Interpretation

- **0–15**: High risk – Immediate action required to address significant gaps in aligning AI investments with business problems.
- **16–30**: Moderate risk – Areas for improvement identified; a structured plan is needed to better align AI investments.
- **31–45**: Low risk – AI investments are well-aligned with business problems, but continuous monitoring and minor adjustments are recommended.

Threshold for Passing

Organizations should aim for a minimum score of **31** to ensure that AI investments are adequately aligned with business problems and are set up to deliver significant value.

Summary

Understanding where to allocate AI investments based on the complexity and scope of business problems is crucial for maximizing returns. This tiered approach helps organizations strategically deploy AI technology, ensuring that each level of investment is justified by corresponding improvements in efficiency, productivity, and innovation. By aligning AI initiatives with specific business needs through a well-structured matrix, companies can create a scalable and impactful AI strategy that adapts to evolving business demands and technological advancements.

CHAPTER 18

AI Cybersecurity

As businesses increasingly integrate artificial intelligence (AI) into their operations, securing AI systems against vulnerabilities and threats becomes paramount. This chapter outlines critical security concerns associated with AI systems and discusses strategies to establish a robust security posture to safeguard against these risks.

Understanding Organizational AI Security Posture

To manage the data needs of multiple teams effectively, a robust AI security posture is essential. This comprehensive approach includes

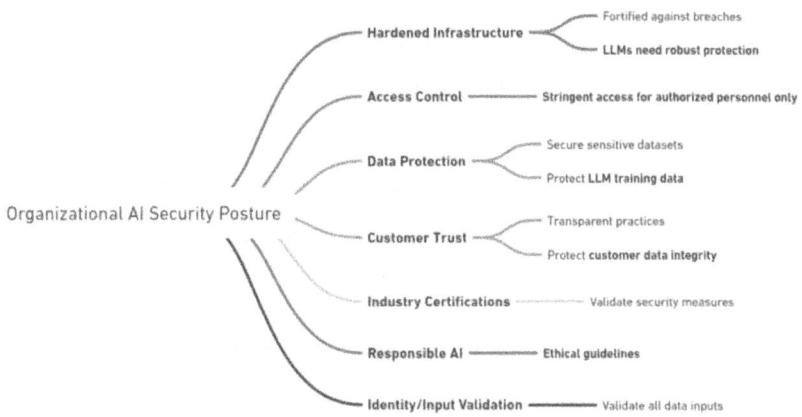

- **Hardened Infrastructure**: Ensuring that all physical and virtual components of AI systems are fortified against breaches. The infrastructure for deploying Large Language Models (LLMs) must be particularly robust as these models often contain billions of parameters that could be exploited if not properly secured.

- **Access Control**: Implementing stringent access controls to ensure that only authorized personnel can interact with the AI systems. Given the complexity and potential power of LLMs, ensuring that access is tightly controlled is paramount.

- **Data Protection**: Employing advanced data protection measures to secure sensitive data from unauthorized access and breaches. In the context of LLMs, this includes securing the massive datasets used for training these models, which often contain sensitive or proprietary information.

- **Customer Trust**: Maintaining customer trust through transparent practices and ensuring the integrity of customer data. This is crucial as LLMs can process a wide range of personal and sensitive data.

- **Industry Certifications**: Obtaining relevant industry certifications that validate the security measures in place.

- **Responsible AI**: Incorporating ethical guidelines and practices to ensure AI is used responsibly.

- **Identity and Input Validation**: Ensuring that all data inputs into AI systems are validated to prevent malicious data from compromising the systems.

Core LLM Security Concerns

Addressing the security concerns specific to Large Language Models (LLMs) is critical for maintaining the integrity and effectiveness of AI systems. The key security issues include

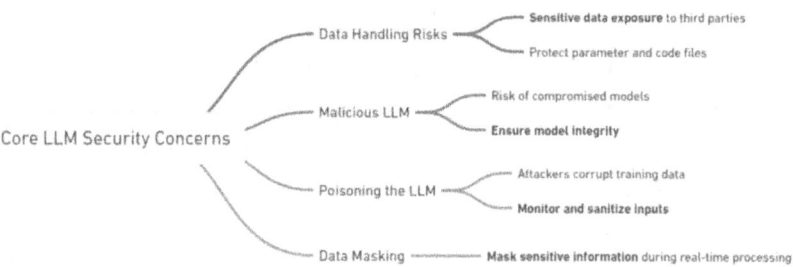

Data Handling Risks: Users may inadvertently share sensitive or proprietary information with third-party LLM providers, leading to potential data leaks. As LLMs can be as simple as two files in a hypothetical directory – a parameter file and a code file – protecting these components is crucial.

- **Malicious LLM**: There are risks of using LLMs that have been compromised to produce harmful outputs or spread misinformation. Ensuring the integrity of the models and their sources is essential.

- **Poisoning the LLM**: Attackers may attempt to corrupt the training data of an LLM, leading to inaccurate or harmful outputs. Monitoring and sanitizing input data during the training phases – pre-training, fine-tuning, and comparison labeling – is vital.

- **Data Masking**: Sensitive information must be adequately masked to prevent unauthorized exposure, especially during model inference, where data is processed in real time.

Strategies to Address AI Security Concerns

Implementing effective strategies and guardrails is crucial for mitigating the risks associated with AI systems:

- **Enhanced Data Handling Protocols**: Establish protocols to control what data is shared with external systems and to monitor data leakage continuously. The management of data across different stages of LLM training – from Internet text aggregation to fine-tuning with proprietary data – needs strict oversight.

- **Secure LLM Integration**: Ensure that any LLM integration passes rigorous security assessments to mitigate the risk of introducing compromised AI into the environment.

- **Robust Poisoning Detection**: Implement advanced analytics to detect anomalies in AI behavior that may indicate a poisoning attack, and set up regular retraining schedules.

- **Advanced Data Masking Techniques**: Utilize state-of-the-art data masking techniques to protect personal and sensitive data from being exposed both internally and externally.

CHAPTER 18   AI CYBERSECURITY

Building a Resilient AI Cybersecurity Framework

To build a resilient AI cybersecurity framework, organizations should focus on

- **Continuous Monitoring and Testing**: Regularly test AI systems for vulnerabilities and monitor them for signs of malicious activity.

- **Training and Awareness**: Conduct regular training sessions for all employees to recognize security threats related to AI and enforce security best practices.

- **Incident Response Planning**: Develop and regularly update an incident response plan that includes procedures for responding to AI-related security breaches.

- **Collaboration with Security Experts**: Work with cybersecurity experts to stay updated on the latest threats and to develop sophisticated defense mechanisms.

CHAPTER 18   AI CYBERSECURITY

AI Cybersecurity Checklist for Board and C-Suite

This checklist is designed to help boards and C-suite executives evaluate and ensure the security of their AI systems. Each question should be scored based on the current state of the organization's AI cybersecurity practices.

Scoring System

- **0 points:** Not addressed
- **1 point:** Partially addressed
- **2 points:** Fully addressed, but needs improvement
- **3 points:** Fully addressed and well-executed

Hardened Infrastructure

1. **Physical and Virtual Security**
   - Are all physical and virtual components of AI systems fortified against breaches?
   - Score: 0-3

2. **LLM Security**
   - Is the infrastructure for deploying Large Language Models (LLMs) robust enough to protect their extensive parameters?
   - Score: 0-3

Access Control

3. **Authorization Management**
   - Are stringent access controls in place to ensure only authorized personnel can interact with the AI systems?
   - Score: 0-3

4. **Access Logging**
   - Are access logs maintained and regularly reviewed for unauthorized access attempts?
   - Score: 0-3

Data Protection

5. **Data Encryption**
   - Is data encryption applied to both data at rest and data in transit to secure sensitive information?
   - Score: 0-3

6. **Training Data Security**
   - Are advanced data protection measures employed to secure the datasets used for training AI models?
   - Score: 0-3

CHAPTER 18   AI CYBERSECURITY

Customer Trust and Transparency

7. **Transparent Practices**
   - Are transparent practices in place to maintain customer trust and ensure the integrity of customer data?
   - Score: 0-3

8. **Industry Certifications**
   - Has the organization obtained relevant industry certifications validating its security measures?
   - Score: 0-3

Responsible AI

9. **Ethical Guidelines**
   - Are ethical guidelines incorporated to ensure AI is used responsibly?
   - Score: 0-3

10. **Bias Mitigation**
    - Are steps taken to mitigate biases in AI systems to ensure fair and ethical use?
    - Score: 0-3

Identity and Input Validation

11. **Data Input Validation**
    - Are all data inputs into AI systems validated to prevent malicious data from compromising the systems?
    - Score: 0-3

12. **Identity Verification**
    - Are robust identity verification mechanisms in place for accessing AI systems?
    - Score: 0-3

Data Handling Risks

13. **Sensitive Data Management**
    - Are protocols established to control what data is shared with external systems and to monitor data leakage continuously?
    - Score: 0-3

14. **Data Masking**
    - Are advanced data masking techniques utilized to protect personal and sensitive data?
    - Score: 0-3

CHAPTER 18 AI CYBERSECURITY

Security of AI Models

15. **Model Integrity**
    - Are security assessments conducted to ensure the integrity of AI models?
    - Score: 0-3

16. **Poisoning Detection**
    - Are advanced analytics implemented to detect anomalies that may indicate a poisoning attack?
    - Score: 0-3

Continuous Monitoring and Testing

17. **Vulnerability Testing**
    - Are AI systems regularly tested for vulnerabilities?
    - Score: 0-3

18. **Activity Monitoring**
    - Are AI systems continuously monitored for signs of malicious activity?
    - Score: 0-3

Training and Awareness

19. **Employee Training**
    - Are regular training sessions conducted for employees to recognize security threats related to AI?
    - Score: 0-3

20. **Best Practices Enforcement**
    - Are security best practices enforced across the organization?
    - Score: 0-3

Incident Response Planning

21. **Incident Response Plan**
    - Is there a comprehensive incident response plan in place for AI-related security breaches?
    - Score: 0-3

22. **Plan Updates**
    - Is the incident response plan regularly updated to address new AI security threats?
    - Score: 0-3

Collaboration with Security Experts

23. **Expert Collaboration**
    - Does the organization collaborate with cybersecurity experts to stay updated on the latest threats?
    - Score: 0-3

24. **Sophisticated Defense Mechanisms**
    - Are sophisticated defense mechanisms developed with the help of security experts?
    - Score: 0-3

CHAPTER 18   AI CYBERSECURITY

Total Scoring

Each question is scored from 0 to 3, with a maximum possible score of 72 across all 24 questions.

Scoring and Interpretation

- **0–24**: High risk – Immediate action required to address significant gaps in AI cybersecurity strategies.
- **25–48**: Moderate risk – Areas for improvement identified, and a structured plan is needed to enhance AI cybersecurity approaches.
- **49–72**: Low risk – Robust AI cybersecurity strategies in place, but continuous monitoring and minor adjustments are recommended.

Threshold for Passing

Organizations should aim for a minimum score of **48** to ensure they have adequately addressed key AI cybersecurity considerations and are prepared to manage and mitigate AI-related risks effectively.

Summary

Establishing a comprehensive cybersecurity posture for AI systems is essential to protect sensitive data, maintain customer trust, and ensure the integrity of AI operations. By addressing the unique challenges posed by AI technology and implementing robust security measures, organizations can mitigate risks and leverage the full potential of AI with confidence. This proactive approach to AI cybersecurity will enable businesses to thrive in an increasingly AI-driven world.

CHAPTER 19

Scaling AI Operations: Designing Effective Enterprise Infrastructure

When embarking on AI initiatives, particularly in enterprise environments, the foundation laid by the underlying architecture and infrastructure is crucial. This chapter explores the essential components and investments needed to establish a robust AI framework capable of supporting the complex needs of modern businesses.

CHAPTER 19 COMPOSING SCALING AI OPERATIONS: DESIGNING EFFECTIVE ENTERPRISE INFRASTRUCTURE

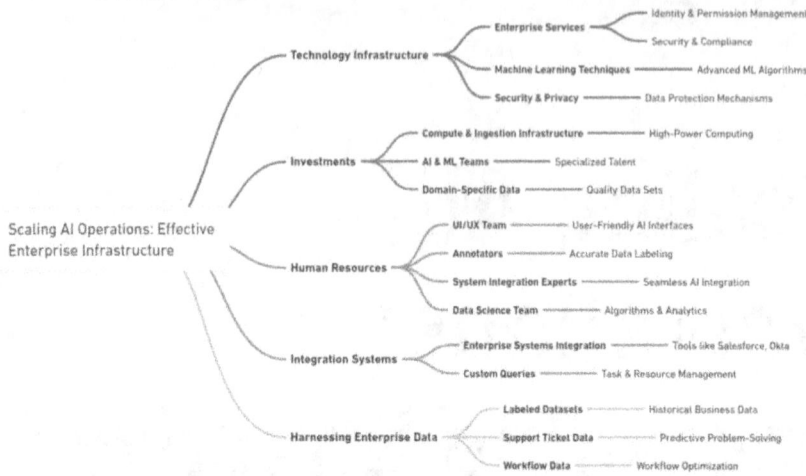

Technology Infrastructure for AI

To kickstart AI deployment, a solid technological base comprising several key components is essential:

- **Enterprise Services:** Critical for identity and permission management, security, compliance, and analytics. These services ensure that AI solutions are seamlessly integrated into the existing enterprise ecosystem, enhancing capabilities without compromising security or performance.

- **Machine Learning Techniques:** The ability to implement advanced ML techniques like reasoning algorithms is fundamental. These techniques are necessary to handle the sophisticated tasks that AI models are expected to perform.

- **Security and Privacy:** Robust mechanisms for data masking, third-party application integrations, and access management are mandatory to protect sensitive information and maintain user trust.

CHAPTER 19   COMPOSING SCALING AI OPERATIONS: DESIGNING EFFECTIVE ENTERPRISE INFRASTRUCTURE

Investments Required to Support AI Initiatives

Launching and maintaining AI capabilities require not only technological setups but also significant investment in several areas:

- **Fine-Tuning Compute and Ingestion Infrastructure:** Powerful computing resources are needed to train, fine-tune, and deploy AI models, especially to handle the large volumes of data typical in enterprise settings.

- **Dedicated AI and ML Teams:** Specialist teams are crucial to drive AI projects from conception through to implementation and ongoing management.

- **Domain-Specific Data:** AI models are as good as the data they learn from. Investing in rich, domain-specific datasets is essential for training effective models.

Human Resources and Expertise

The success of AI projects also hinges on having the right talent and expertise:

- **UI/UX Team:** Ensuring that AI interfaces are user-friendly and efficient requires skilled UI/UX professionals.

- **Annotators:** These teams are responsible for labeling data accurately, which is crucial for training AI models.

- **System Integration Experts:** These professionals ensure that AI systems are correctly integrated with existing IT infrastructure.

- **Data Science Team:** This team develops the algorithms and analytics that power AI applications, turning raw data into actionable insights.

Integration Systems for Expanded AI Scope

As AI applications scale, the ability to integrate with various enterprise systems becomes vital:

- **Common Enterprise Systems:** Tools like Salesforce, Okta, Calendly, and Azure AD need to be integrated seamlessly to leverage AI across business functions.

- **Custom Requests and Queries:** AI systems should be capable of handling specific requests such as scheduling meetings, managing tasks, or locating resources within the enterprise.

Harnessing Unique, Task-Specific Enterprise Data

The utility of an AI system is greatly enhanced by its access to unique, labeled enterprise data:

- **Enterprise-Labeled Dataset:** Utilizing historical data accumulated over years provides AI models with the depth of knowledge required to perform effectively.

- **Support Ticket Data:** By analyzing past support tickets, AI can learn to predict and resolve future issues more efficiently.

- **Workflow Data:** Understanding the steps in business processes helps AI to optimize workflows and suggest improvements.

Checklist for "Scaling AI Operations: Designing Effective Enterprise Infrastructure"

This checklist is designed for board members and C-suite executives to assess the readiness and effectiveness of their organization's AI infrastructure. It covers critical areas that must be considered to ensure successful scaling of AI operations. The scoring system is quantitative, with each question assigned a score based on the organization's current practices.

Scoring System

- **0 points:** Not addressed
- **1 point:** Partially addressed
- **2 points:** Fully addressed, but needs improvement
- **3 points:** Fully addressed and well-executed

1. Technology Infrastructure for AI

1. **Enterprise Services Integration** (Score: 0-3)
   - Are enterprise services like identity and permission management, security, compliance, and analytics seamlessly integrated into the AI infrastructure?

2. **Advanced Machine Learning Techniques**
   (Score: 0-3)

   - Does the AI infrastructure support the implementation of advanced ML techniques, such as reasoning algorithms, to handle sophisticated tasks?

3. **Security and Privacy Mechanisms** (Score: 0-3)

   - Are robust security and privacy measures, including data masking, third-party application integrations, and access management, in place to protect sensitive information?

2. Investments Required to Support AI Initiatives

4. **Compute and Ingestion Infrastructure** (Score: 0-3)

   - Are powerful computing resources available to train, fine-tune, and deploy AI models, particularly for handling large volumes of data?

5. **Dedicated AI and ML Teams** (Score: 0-3)

   - Has the organization invested in specialist teams dedicated to driving AI projects from conception to implementation and ongoing management?

6. **Domain-Specific Data Investment** (Score: 0-3)

   - Is there significant investment in acquiring and managing rich, domain-specific datasets necessary for training effective AI models?

3. Human Resources and Expertise

7. **UI/UX Team Competence** (Score: 0-3)

   - Are there skilled UI/UX professionals in place to ensure that AI interfaces are user-friendly and efficient?

8. **Annotators for Data Labeling** (Score: 0-3)

   - Does the organization have dedicated annotators responsible for accurately labeling data, which is crucial for training AI models?

9. **System Integration Experts** (Score: 0-3)

   - Are there professionals tasked with ensuring that AI systems are correctly integrated with existing IT infrastructure?

10. **Data Science Team Capabilities** (Score: 0-3)

    - Does the organization have a competent data science team to develop algorithms and analytics that power AI applications?

4. Integration Systems for Expanded AI Scope

11. **Integration with Common Enterprise Systems** (Score: 0-3)

    - Are AI systems integrated seamlessly with common enterprise tools like Salesforce, Okta, Calendly, and Azure AD to leverage AI across business functions?

12. **Handling Custom Requests and Queries**
    (Score: 0-3)

    - Are AI systems capable of handling custom requests, such as scheduling meetings, managing tasks, or locating resources within the enterprise?

5. Harnessing Unique, Task-Specific Enterprise Data

13. **Utilization of Enterprise-Labeled Datasets**
    (Score: 0-3)

    - Is the organization effectively utilizing historical data accumulated over years to provide AI models with the depth of knowledge required for optimal performance?

14. **Analysis of Support Ticket Data** (Score: 0-3)

    - Does the AI system analyze past support tickets to predict and resolve future issues more efficiently?

15. **Optimization of Workflow Data** (Score: 0-3)

    - Is workflow data being utilized to help AI optimize business processes and suggest improvements?

Total Scoring

Each question is scored from 0 to 3, with a maximum possible score of 45 across all 15 questions.

Scoring and Interpretation

- **0–15**: High risk – Immediate action required to address significant gaps in AI infrastructure and operations.

- **16–30**: Moderate risk – Areas for improvement identified, and a structured plan is needed to enhance AI infrastructure.

- **31–45**: Low risk – Robust AI infrastructure is in place, but continuous monitoring and minor adjustments are recommended.

Threshold for Passing

Organizations should aim for a minimum score of **30** to ensure they have adequately addressed key considerations for scaling AI operations and are prepared to support AI deployment effectively at an enterprise level.

As AI moves from buzzword to business imperative, the scaffolding that supports it becomes the linchpin of success – is your enterprise infrastructure ready for the AI revolution? From robust technology stacks to specialized human expertise, scaling AI operations demands a holistic approach that marries cutting-edge tech with deep domain knowledge. The integration of AI into enterprise ecosystems isn't just about powerful algorithms; it's about creating a symbiotic relationship between machines and existing business processes, all while navigating the complex terrain of security, privacy, and compliance. As organizations stand at the crossroads of AI adoption, those who invest wisely in infrastructure, talent, and data will find themselves not just participants in the AI race, but frontrunners poised to redefine their industries – will your enterprise infrastructure be the launchpad for AI innovation, or the bottleneck that holds you back?

CHAPTER 20

Building Robust AI Infrastructure for Enterprise Success

When embarking on AI initiatives, particularly in enterprise environments, the foundation laid by the underlying architecture and infrastructure is crucial. This chapter explores the essential components and investments needed to establish a robust AI framework capable of supporting the complex needs of modern businesses.

What Are Large Language Models?

LLMs are essentially large neural networks that process and generate human language. They consist of two main components: a parameter file and a neural network. The parameter file contains the model's weights, which are used to make predictions, while the neural network processes the input data. LLMs can be thought of as "Zip files of the internet," containing a vast amount of knowledge and information.

CHAPTER 20 BUILDING ROBUST AI INFRASTRUCTURE FOR ENTERPRISE SUCCESS

Model Architecture and Training

The transformer architecture is the underlying framework for most LLMs. This architecture allows the model to process input sequences of arbitrary length and generate output sequences of similar length. The model is trained on a massive dataset of text, enabling it to learn the patterns and relationships between words and phrases.

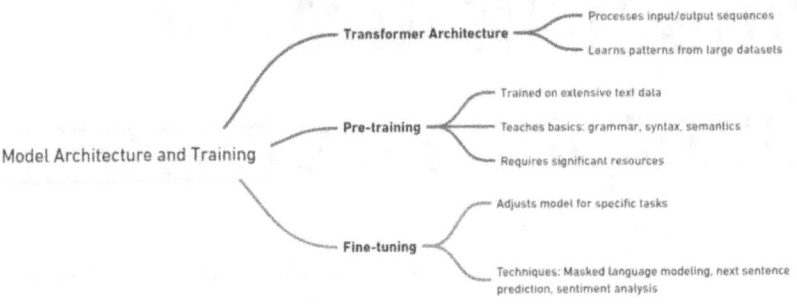

- **Pre-training and Fine-tuning**

    - **Pre-training:** Training the model on a large corpus of text data, which can be Internet documents, books, or other sources of text. This stage teaches the model the basics of language, including grammar, syntax, and semantics. The pre-training process can be computationally expensive, requiring significant resources and infrastructure.

    - **Fine-tuning:** Adjusting the model's parameters to fit a specific task or dataset. This stage is crucial in adapting the model to a particular application or use case. Fine-tuning can be done using various techniques, including masked language modeling, next sentence prediction, and sentiment analysis.

LLM Security and Risks

LLMs are not immune to security risks and challenges. One primary concern is data poisoning, where malicious actors inject harmful data into the training dataset, resulting in the model generating biased or toxic output. Another risk is prompt injection attacks, where attackers manipulate the input prompts to elicit specific responses from the model.

LLM Applications

LLMs have numerous applications in various industries, including but not limited to

- **Natural Language Processing (NLP):** LLMs can be used for text classification, sentiment analysis, and language translation.
- **Conversational AI:** LLMs generate human-like responses to user queries, making them suitable for chatbots and virtual assistants.
- **Content Generation:** LLMs can create high-quality content, such as articles, stories, and even entire books.
- **Multimodal Applications:** LLMs can be used for image-to-text and voice-to-text tasks, enabling applications like image captioning and speech-to-text systems.

Optimizing LLM Performance

Optimizing LLM performance involves several key factors:

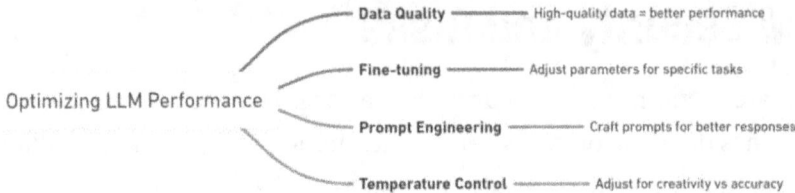

- **Data Quality:** The quality of the training data significantly affects the model's performance. High-quality data leads to better results.

- **Fine-tuning:** Adjusting the model's parameters to fit a specific task or dataset can significantly improve performance.

- **Prompt Engineering:** Crafting well-designed prompts can elicit more accurate and informative responses from the model.

- **Temperature Control:** Adjusting the model's temperature can influence the level of creativity or accuracy in the generated output.

LLM Evaluation Metrics

Evaluating LLM performance is crucial in understanding their capabilities and limitations. Some common evaluation metrics include

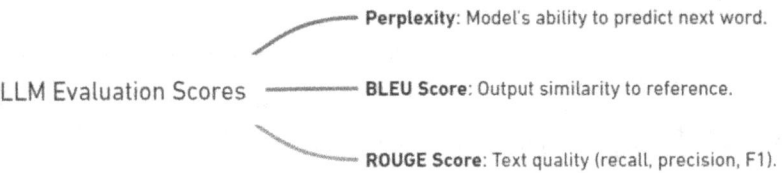

- **Perplexity:** Measures the model's ability to predict the next word in a sequence.

- **BLEU Score:** Measures the similarity between the model's output and the reference output.

- **ROUGE Score:** Measures the quality of the generated text based on recall, precision, and F1 score.

Challenges and Future Directions

While LLMs have achieved remarkable success, they also face challenges and limitations. Key concerns include

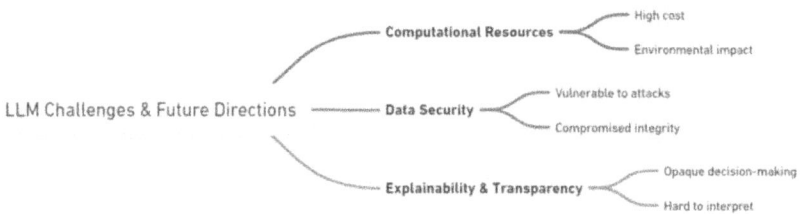

- **Computational Resources:** Training and deploying LLMs require significant computational resources, which can be costly and environmentally unsustainable.

- **Data Security:** LLMs can be vulnerable to data poisoning and backdoor attacks, compromising their performance and integrity.

- **Explainability and Transparency:** LLMs can be opaque and difficult to interpret, making it challenging to understand their decision-making processes.

CHAPTER 20   BUILDING ROBUST AI INFRASTRUCTURE FOR ENTERPRISE SUCCESS

Checklist for "Building Robust AI Infrastructure for AI Success"

This checklist is designed for board members and C-suite executives to assess the readiness and effectiveness of their organization's AI infrastructure. The checklist is divided into key areas critical to building and sustaining a robust AI infrastructure. Each question is scored individually, and organizations should aim for a minimum threshold to ensure they are adequately prepared to deploy and scale AI initiatives.

Scoring System

- **0 points:** Not addressed
- **1 point:** Partially addressed
- **2 points:** Fully addressed, but needs improvement
- **3 points:** Fully addressed and well-executed

Technology Infrastructure for AI

1. **Enterprise Services**
   - Have we implemented enterprise services for identity and permission management, security, compliance, and analytics?
   - **Score:** ____/3

2. **Machine Learning Techniques**
   - Are advanced machine learning techniques, such as reasoning algorithms, effectively implemented to handle sophisticated AI tasks?
   - **Score:** ____/3

3. **Security and Privacy**

    - Are robust security mechanisms in place, including data masking, third-party application integrations, and access management, to protect sensitive information?

    - Score: ____/3

Investments Required to Support AI Initiatives

4. **Compute and Ingestion Infrastructure**

    - Do we have sufficient and scalable computing resources to train, fine-tune, and deploy AI models, especially for handling large data volumes?

    - Score: ____/3

5. **Dedicated AI and ML Teams**

    - Have we established specialist teams dedicated to driving AI projects from conception through implementation and ongoing management?

    - Score: ____/3

6. **Domain-Specific Data**

    - Are we investing in rich, domain-specific datasets necessary for training effective AI models?

    - Score: ____/3

Human Resources and Expertise

7. **UI/UX Team**
   - Is there a skilled UI/UX team in place to ensure that AI interfaces are user-friendly and efficient?
   - Score: ____/3

8. **Annotators**
   - Do we have dedicated annotators responsible for accurately labeling data, which is crucial for training AI models?
   - Score: ____/3

9. **System Integration Experts**
   - Are system integration experts available to ensure that AI systems are correctly integrated with existing IT infrastructure?
   - Score: ____/3

10. **Data Science Team**
    - Is there a capable data science team responsible for developing algorithms and analytics that power AI applications?
    - Score: ____/3

Integration Systems for Expanded AI Scope

11. **Common Enterprise Systems**

    - Have we integrated AI systems with common enterprise tools such as Salesforce, Okta, Calendly, and Azure AD to leverage AI across business functions?

    - Score: ____/3

12. **Custom Requests and Queries**

    - Can our AI systems handle custom requests, such as scheduling meetings, managing tasks, or locating resources within the enterprise?

    - Score: ____/3

Harnessing Unique, Task-Specific Enterprise Data

13. **Enterprise Labeled Dataset**

    - Are we utilizing historical enterprise data that has been accumulated over the years to provide AI models with the depth of knowledge required?

    - Score: ____/3

14. **Support Ticket Data**

    - Is AI being used to analyze past support tickets to predict and resolve future issues more efficiently?

    - Score: ____/3

15. **Workflow Data**

    - Are AI systems leveraging workflow data to understand business processes and optimize them by suggesting improvements?

    - **Score:** ____/3

Technical Understanding of Generative AI for Boards and Executives

16. **LLM Implementation**

    - Do we understand the architecture and training processes behind Large Language Models (LLMs), including their capabilities and limitations?

    - **Score:** ____/3

17. **LLM Security Measures**

    - Have we implemented security measures to protect LLMs from data poisoning, prompt injection attacks, and other security threats?

    - **Score:** ____/3

18. **LLM Performance Optimization**

    - Are strategies in place for optimizing LLM performance, including fine-tuning, prompt engineering, and maintaining data quality?

    - **Score:** ____/3

Total Scoring and Interpretation

Each question is scored from 0 to 3, with a maximum possible score of **54** across all 18 questions.

Scoring Threshold

- **0-18:** High risk – Significant gaps exist; immediate action is required to address AI infrastructure shortcomings.
- **19-36:** Moderate risk – Some areas are well-managed, but there is a need for further development to ensure robust AI infrastructure.
- **37-54:** Low risk – The organization has a well-established AI infrastructure that is ready to support large-scale AI initiatives effectively.

Minimum Score to Pass

Organizations should aim for a minimum score of **36** to ensure they have adequately addressed key aspects of building a robust AI infrastructure. This threshold indicates that while there may be areas for improvement, the organization is generally prepared to support and scale AI operations successfully.

CHAPTER 20   BUILDING ROBUST AI INFRASTRUCTURE FOR ENTERPRISE SUCCESS

Summary

Setting up a robust AI architecture and infrastructure requires careful planning and significant resources, both technological and human. By investing in the right areas and ensuring seamless integration of AI into the existing enterprise fabric, businesses can unlock the full potential of AI to drive innovation and efficiency. This comprehensive approach supports current AI applications and lays the groundwork for future advancements.

CHAPTER 21

Architecting AI Solutions: A Blueprint for Generative AI

The Generative AI Reference Architecture

The Generative AI Reference Architecture is a structured approach to developing AI solutions, encompassing various stages from data preparation to deployment and monitoring. This architecture ensures that AI systems are secure, scalable, and capable of delivering significant business value.

CHAPTER 21 ARCHITECTING AI SOLUTIONS: A BLUEPRINT FOR GENERATIVE AI

1. User Experience (U/X)

The architecture begins with the user experience, focusing on building intuitive interfaces and conversational agents. A rich developer and workplace experience is crucial for ensuring that AI solutions are user-friendly and meet the needs of end users.

- **Search Enterprise Data**: Efficient data retrieval from enterprise databases is essential for building effective AI models.

- **Conversational Agents**: Developing conversational interfaces enhances user interaction and accessibility.

2. Prompt Engineering

Prompt engineering involves designing prompts that elicit the desired responses from AI models. This stage uses advanced prompting techniques to refine model outputs.

- **Re-engineer Prompts**: Use basic to advanced prompting techniques to produce accurate and relevant outputs.

- **Prompt Best Practices**: Implement techniques such as ICL (In-Context Learning), CoT (Chain of Thought), and ReAct (Reflective Action) to enhance prompt effectiveness.

3. Retrieval Augmentation Generation (RAG)

RAG techniques are used to retrieve additional data that augments the prompt, enhancing the model's performance.

Data Sources: Utilize unstructured data, vector databases, SQL, and graph databases for comprehensive data augmentation.

4. Serving and Orchestrating Models

Generating and coordinating model outputs is a critical aspect of the architecture. This involves managing the flow of data and orchestrating multiagent systems to perform complex tasks.

- **Model Garden and Registry**: Maintain a repository of AI models to ensure easy access and management.

- **Agent Flow**: Route and orchestrate tasks across various AI agents for optimal performance.

5. Adaptation and Tuning

Adapting and tuning models to specific tasks and datasets is crucial for achieving high performance and accuracy.

- **Extensions and Connectors**: Use extensions, distillation methods, and function calling to enhance model capabilities.
- **Tuning Techniques**: Employ techniques such as synthetic data generation and domain-specific customization to refine models.

6. Evaluation and Observability

Continuous evaluation and observability are necessary to maintain the reliability and effectiveness of AI systems.

- **Ground Truth**: Validate AI outputs against a reliable source of truth to ensure accuracy.
- **Observability Frameworks**: Implement frameworks to monitor AI system performance and identify issues in real time.

7. MLOps Orchestration

Machine Learning Operations (MLOps) involves integrating and monitoring AI models throughout their lifecycle, from development to deployment and maintenance.

- **Data Pipeline**: Create pipelines that integrate and monitor data, ensuring seamless operation and continuous improvement of AI systems.

- **Predictive and Generative AI**: Leverage both predictive and generative AI capabilities to address diverse business needs.

8. Security, Privacy, and Compliance

Ensuring the security, privacy, and compliance of AI systems is paramount. This involves safeguarding data, code, and models against unauthorized access and ensuring adherence to regulatory standards.

- **Secure Data**: Implement robust security measures to protect sensitive data.

- **Compliance Checks**: Regularly conduct compliance checks to ensure adherence to legal and ethical standards.

9. Governance and Responsible AI

Responsible AI practices are integral to building trustworthy AI systems. This includes implementing unbiased safety checks and recitation checks to maintain fairness and transparency.

- **Unbiased Safety Checks**: Ensure that AI systems operate without bias and produce fair outcomes.

- **Recitation Checks**: Validate the correctness and reliability of AI-generated content.

10. Enterprise Integration

Integrating AI systems with enterprise backend systems is essential for seamless operation and scalability.

Backend Systems: Ensure that AI models are compatible with existing enterprise infrastructure for smooth integration.

Industry-Specific Customization: Tailor AI solutions to meet the specific needs of different industry domains.

Strategic Implications for Businesses

Implementing the Generative AI Reference Architecture offers several strategic benefits for businesses:

1. Enhanced Innovation

By leveraging advanced AI techniques and architectures, businesses can drive significant innovation, developing new products and services that meet evolving market demands.

2. Operational Efficiency

AI systems can streamline operations, reduce manual intervention, and enhance productivity, leading to significant cost savings and efficiency gains.

3. Competitive Advantage

Adopting a robust AI architecture enables businesses to stay ahead of competitors by quickly adapting to technological advancements and market changes.

4. Risk Mitigation

Implementing comprehensive security, privacy, and compliance measures ensures that AI systems operate within legal and ethical boundaries, reducing the risk of regulatory penalties and reputational damage.

5. Scalability

A well-architected AI system can scale effectively, accommodating growing data volumes and increasing user demands without compromising performance.

Building Trustworthy AI Systems

As AI technologies continue to integrate into various facets of business operations, ensuring the trustworthiness of these systems becomes paramount. This chapter outlines the critical characteristics that define trustworthy AI systems, providing a comprehensive guide for corporate boards to oversee and ensure the responsible implementation of AI within their organizations.

CHAPTER 21   ARCHITECTING AI SOLUTIONS: A BLUEPRINT FOR GENERATIVE AI

Characteristics of Trustworthy AI Systems

A trustworthy AI system is defined by several key characteristics that collectively ensure its safe, ethical, and effective use. These characteristics include safety, security, explainability, privacy, fairness, accountability, and reliability.

1. **Safe AI Systems** – Safety is the cornerstone of trustworthy AI. AI systems must be designed to operate safely within their intended environments, minimizing the risk of harm to users and stakeholders.

   Implementation: Safety protocols and fail-safes must be integrated into AI systems to handle unexpected scenarios and mitigate potential risks.

2. **Secure and Resilient AI Systems** – Security and resilience are crucial for protecting AI systems from cyber threats and ensuring their continued operation under adverse conditions.

   Implementation: Employ robust encryption methods, regular security audits, and resilience testing to safeguard AI systems against breaches and disruptions.

3. **Explainable and Interpretable AI Systems** – For AI to be trusted, it must be explainable and interpretable. Stakeholders need to understand how AI systems make decisions to ensure transparency and accountability.

   Implementation: Develop AI models with inherent explainability features and use interpretability tools to provide insights into AI decision-making processes.

4. **Privacy-Enhanced AI Systems** – AI systems must enhance privacy by protecting personal data and complying with relevant data protection regulations.

   Implementation: Utilize anonymization, encryption, and differential privacy techniques to secure personal data and ensure compliance with laws like GDPR and HIPAA.

5. **Fair AI Systems with Managed Bias** – Fairness in AI involves addressing and mitigating harmful biases in AI models to ensure equitable outcomes for all users.

   Implementation: Conduct regular bias audits, use diverse training datasets, and implement fairness algorithms to detect and correct biases in AI systems.

6. **Accountable and Transparent AI Systems** – Accountability and transparency are vital for building trust in AI systems. Organizations must take responsibility for AI outcomes and be transparent about AI use.

    Implementation: Establish clear accountability frameworks, document AI processes, and provide stakeholders with access to information about AI system operations.

7. **Valid and Reliable AI Systems** – The foundation of a trustworthy AI system is its validity and reliability. AI systems must consistently perform as intended and deliver accurate results.

    Implementation: Rigorously test AI models, validate their outputs, and ensure they remain reliable over time through continuous monitoring and updates.

Checklist for Architecting AI Solutions: A Blueprint for Generative AI

This checklist is designed for board members and C-suite executives to ensure the successful implementation of AI solutions. Each question is scored individually to provide a comprehensive assessment of readiness and maturity in AI solution architecture.

Scoring System

- **0 points:** Not addressed
- **1 point:** Partially addressed
- **2 points:** Fully addressed, but needs improvement
- **3 points:** Fully addressed and well-executed

User Experience (UX)

1. **Search Enterprise Data** (Score: 0-3)
   - Are the AI solutions integrated with enterprise databases to ensure efficient data retrieval?
2. **Conversational Agents** (Score: 0-3)
   - Have user-friendly conversational interfaces been developed to enhance interaction and accessibility?

Prompt Engineering

3. **Re-engineer Prompts** (Score: 0-3)
   - Are advanced prompting techniques used to produce accurate and relevant outputs?
4. **Prompt Best Practices** (Score: 0-3)
   - Are techniques such as In-Context Learning (ICL), Chain of Thought (CoT), and Reflective Action (ReAct) implemented to enhance prompt effectiveness?

CHAPTER 21 ARCHITECTING AI SOLUTIONS: A BLUEPRINT FOR GENERATIVE AI

Retrieval Augmentation Generation (RAG)

5. **Data Sources** (Score: 0-3)
   - Are various data sources like unstructured data, vector databases, SQL, and graph databases utilized for comprehensive data augmentation?

Serving and Orchestrating Models

6. **Model Garden and Registry** (Score: 0-3)
   - Is there a repository of AI models to ensure easy access and management?

7. **Agent Flow** (Score: 0-3)
   - Are tasks routed and orchestrated across various AI agents for optimal performance?

Adaptation and Tuning

8. **Extensions and Connectors** (Score: 0-3)
   - Are extensions, distillation methods, and function calling used to enhance model capabilities?

9. **Tuning Techniques** (Score: 0-3)
   - Are techniques such as synthetic data generation and domain-specific customization employed to refine models?

CHAPTER 21   ARCHITECTING AI SOLUTIONS: A BLUEPRINT FOR GENERATIVE AI

Evaluation and Observability

10. **Ground Truth** (Score: 0-3)

    - Are AI outputs validated against a reliable source of truth to ensure accuracy?

11. **Observability Frameworks** (Score: 0-3)

    - Are frameworks implemented to monitor AI system performance and identify issues in real time?

MLOps Orchestration

12. **Data Pipeline** (Score: 0-3)

    - Are pipelines created to integrate and monitor data, ensuring seamless operation and continuous improvement of AI systems?

13. **Predictive and Generative AI** (Score: 0-3)

    - Are both predictive and generative AI capabilities leveraged to address diverse business needs?

Security, Privacy, and Compliance

14. **Secure Data** (Score: 0-3)

    - Are robust security measures implemented to protect sensitive data?

15. **Compliance Checks** (Score: 0-3)

    - Are regular compliance checks conducted to ensure adherence to legal and ethical standards?

CHAPTER 21   ARCHITECTING AI SOLUTIONS: A BLUEPRINT FOR GENERATIVE AI

Governance and Responsible AI

16. **Unbiased Safety Checks** (Score: 0-3)
    - Are unbiased safety checks in place to ensure that AI systems operate without bias and produce fair outcomes?

17. **Recitation Checks** (Score: 0-3)
    - Is the correctness and reliability of AI-generated content validated through recitation checks?

Enterprise Integration

18. **Backend Systems** (Score: 0-3)
    - Are AI models compatible with existing enterprise infrastructure for smooth integration?

19. **Industry-Specific Customization** (Score: 0-3)
    - Are AI solutions tailored to meet the specific needs of different industry domains?

Total Scoring

Each question is scored from 0 to 3, with a maximum possible score of 57 across all 19 questions.

Scoring and Interpretation

- **0-19:** High risk – Immediate action required to address significant gaps in AI solution architecture.

- **20–38:** Moderate risk – Areas for improvement identified, and a structured plan is needed to enhance AI solution architecture.

- **39–57:** Low risk – Robust AI solution architecture in place, but continuous monitoring and minor adjustments are recommended.

Threshold for Passing

Organizations should aim for a minimum score of **38** to ensure they have adequately addressed key considerations for AI solution architecture and are prepared to implement and scale AI operations effectively.

Summary

In the grand chessboard of AI implementation, a well-architected solution isn't just a nice-to-have – it's the kingmaker that can crown your business as an industry leader or leave you in digital checkmate. From user experience to enterprise integration, the Generative AI Reference Architecture provides a comprehensive blueprint for building AI systems that are not just powerful, but trustworthy, scalable, and aligned with business objectives. As AI becomes the linchpin of innovation and competitive advantage, the imperative for boards and C-suites is clear: architect your AI solutions with the same meticulous care you'd apply to your most critical business strategies. In this brave new world, where AI isn't just augmenting human intelligence but reshaping entire industries, will your AI architecture be the fortress that propels you to market dominance, or the Achilles' heel that leaves you vulnerable to more nimble competitors?

CHAPTER 22

AI Risk Categorization

While AI can enhance efficiency, drive innovation, and provide deep insights, it also introduces new challenges that businesses must address. Understanding and mitigating these risks is essential for any organization seeking to leverage AI effectively. This chapter explores the various risks associated with AI, from operational and legal to ethical concerns, and provides strategies for managing these risks.

Operational Risks

Operational risks in AI arise from the complexity and unpredictability of AI systems. These risks can impact business continuity and performance if not properly managed.

CHAPTER 22 AI RISK CATEGORIZATION

1. Unpredictable AI Behavior

AI systems, especially those based on machine learning, can exhibit unpredictable behavior due to the complexity of their algorithms and the vast amount of data they process. This unpredictability can lead to unintended consequences, such as biased decisions or operational failures.

2. Data Quality and Integrity

The effectiveness of AI heavily depends on the quality of the data it is trained on. Poor data quality can lead to inaccurate predictions and flawed decision-making. Ensuring data integrity is crucial for maintaining the reliability of AI systems.

3. Dependency on External Vendors

Relying on third-party AI solutions can expose businesses to additional risks, such as vendor lock-in, lack of transparency, and potential disruptions in service. It is important to have clear agreements and contingency plans in place to manage these dependencies.

Legal and Regulatory Risks

Legal and regulatory risks are significant when deploying AI, as the legal framework surrounding AI is still evolving. Businesses must navigate these uncertainties to avoid legal pitfalls.

1. Liability and Accountability

Determining liability for decisions made by AI systems can be complex. If an AI system makes a faulty decision that causes harm, it can be unclear whether the responsibility lies with the AI developer, the vendor, or the user organization. Clear legal frameworks and accountability measures are essential.

2. Data Privacy and Protection

AI systems often require vast amounts of personal data, raising concerns about data privacy and protection. Compliance with regulations such as GDPR and CCPA is critical to avoid legal repercussions and maintain customer trust.

3. Compliance with Industry Standards

Different industries have specific regulations that govern the use of AI. Ensuring compliance with these standards is necessary to avoid penalties and legal challenges.

Ethical and Social Risks

Ethical and social risks associated with AI involve the impact of AI on society and the ethical considerations in its deployment and use.

1. Bias and Discrimination

AI systems can perpetuate and even amplify biases present in the training data. This can lead to discriminatory practices, particularly in sensitive areas such as hiring, lending, and law enforcement. Addressing bias in AI is crucial to ensure fairness and equity.

2. Transparency and Explainability

Many AI systems operate as "black boxes," making it difficult to understand how decisions are made. Lack of transparency and explainability can undermine trust and lead to ethical concerns. Developing explainable AI models is essential for accountability.

3. Impact on Employment

AI has the potential to disrupt labor markets by automating tasks previously performed by humans. This can lead to job displacement and require strategies for workforce reskilling and transition.

Checklist for AI Risk Consideration for Board and C-Suite
Privacy

1. **Data Collection and Usage**
   - Are we collecting only the necessary data for AI applications?
   - How is the data being anonymized to protect individual privacy?
   - Are there robust data minimization strategies in place?

2. **Data Subject Rights**
   - How do we ensure compliance with data subject rights, such as access and deletion requests?
   - What processes are in place to inform individuals about data use?

CHAPTER 22   AI RISK CATEGORIZATION

Security

3. **Model Selection and Training**
   - What measures are taken to secure the AI models from adversarial attacks during training?
   - How are we ensuring the integrity and confidentiality of the training data?

4. **System and Data Security**
   - Are our AI systems and data encrypted and protected from breaches?
   - What protocols are in place for regular security audits and vulnerability assessments?

Fairness

5. **Bias and Discrimination**
   - How do we test and mitigate bias in our AI algorithms?
   - Are diverse datasets being used to train AI models to ensure fairness?

6. **Equity and Inclusivity**
   - How do we ensure that AI decisions are equitable and inclusive?
   - Are there checks to prevent historical biases from influencing AI outcomes?

Transparency and Explainability

7. **Model Interpretability**
   - Can we explain the decision-making process of our AI models to stakeholders?
   - What steps are taken to ensure that AI outcomes are understandable and transparent?

8. **Documentation and Reporting**
   - How are decisions made by AI systems documented?
   - What mechanisms are in place for reporting AI performance and decision rationales?

Safety and Performance

9. **Operational Safety**
   - What measures are in place to ensure AI systems operate safely under various conditions?
   - How do we test AI systems for robustness against failures and unexpected scenarios?

10. **Performance Monitoring**
    - How is the performance of AI systems continuously monitored and evaluated?
    - What are the key performance indicators (KPIs) for AI effectiveness and reliability?

CHAPTER 22   AI RISK CATEGORIZATION

Third-Party Risks

11. **Vendor and Partner Management**
    - How do we assess the risks associated with third-party AI vendors and partners?
    - What criteria are used to select and evaluate third-party AI solutions?

12. **Contractual Obligations**
    - Are there clear contracts outlining the responsibilities and liabilities of third-party vendors?
    - How do we ensure that third-party AI solutions comply with our security and privacy standards?

Legal and Regulatory Compliance

13. **Regulatory Adherence**
    - How do we ensure compliance with relevant AI regulations and standards?
    - Are there processes to stay updated with evolving AI regulations?

14. **Liability and Accountability**
    - What legal frameworks are in place to address potential liabilities from AI use?
    - How is accountability for AI decisions assigned and managed within the organization?

Organizational and Cultural Integration

15. **AI Literacy and Training**
    - What training programs are in place to enhance AI literacy among employees?
    - How do we promote a culture of ethical AI use across the organization?

16. **Governance and Oversight**
    - How is AI governance structured within the organization?
    - What roles and responsibilities are defined for AI oversight and management?

By addressing these questions, boards and C-suite executives can ensure comprehensive risk management in the deployment and use of AI systems, aligning with strategic goals while safeguarding against potential threats.

Strategies for Mitigating AI Risks

Effectively managing AI risks involves a multifaceted approach that includes technical, organizational, and legal strategies.

CHAPTER 22   AI RISK CATEGORIZATION

1. Robust Risk Management Framework

Develop a comprehensive risk management framework that identifies, assesses, and mitigates AI risks. This framework should be integrated into the overall risk management strategy of the organization.

2. Continuous Monitoring and Auditing

Implement continuous monitoring and auditing of AI systems to detect and address issues promptly. Regular audits can help ensure compliance with legal and ethical standards and maintain the integrity of AI systems.

3. Clear Legal Agreements

Establish clear legal agreements with AI vendors and partners that define liability, accountability, and data protection measures. These agreements should include provisions for compliance with relevant regulations and standards.

4. Data Governance and Quality Control

Implement robust data governance practices to ensure the quality and integrity of data used by AI systems. Regularly review and update data sources to eliminate biases and inaccuracies.

5. Transparent and Explainable AI

Develop AI systems that are transparent and explainable. This involves creating models that can be easily understood and interpreted by humans, ensuring accountability and building trust with stakeholders.

6. Ethical AI Guidelines

Adopt ethical AI guidelines that outline the principles and practices for the responsible use of AI. These guidelines should address issues such as bias, transparency, and the impact on employment, and be aligned with the organization's values.

7. Workforce Reskilling and Transition

Prepare for the impact of AI on the workforce by investing in reskilling and upskilling programs. Support employees in transitioning to new roles that leverage AI and ensure they have the skills needed for the future of work.

Case Study: Managing AI Risks in Financial Services

A mid-sized financial services firm sought to implement an AI-powered chatbot to improve customer service and reduce operational costs. While the potential benefits were clear, the firm faced several risks that needed to be managed.

CHAPTER 22   AI RISK CATEGORIZATION

1. Risk Assessment and Mitigation

The firm conducted a thorough risk assessment to identify potential issues, such as data privacy concerns and the risk of biased responses. They implemented data anonymization techniques to protect customer information and used diverse training data to mitigate bias.

2. Legal and Compliance Strategy

To address legal risks, the firm worked closely with legal advisors to ensure compliance with financial regulations and data protection laws. They established clear contracts with the AI vendor that outlined liability and accountability.

3. Continuous Monitoring and Adjustment

The chatbot's performance was continuously monitored to identify and correct any issues promptly. Regular audits were conducted to ensure compliance with ethical and legal standards.

4. Transparency and Customer Trust

To build trust with customers, the firm made the chatbot's decision-making process transparent and provided clear explanations for its responses. They also maintained a human-in-the-loop to handle complex or sensitive inquiries.

CHAPTER 22   AI RISK CATEGORIZATION

Summary

The deployment of AI in business operations offers transformative potential but also introduces significant risks that must be managed proactively. By understanding the various risks associated with AI and implementing robust strategies for mitigation, organizations can harness the power of AI while safeguarding their operations, reputation, and stakeholders. Developing a comprehensive AI risk management framework, ensuring legal and ethical compliance, and fostering transparency and accountability are key to navigating the complexities of AI and achieving sustainable success

CHAPTER 23

Strategic AI Risk Management & Quantification

The framework for AI risk management involves three critical steps: mapping risks, measuring their impact, and managing them effectively. At the core of this process is strong governance, which ensures that risk management practices are embedded within the organizational culture.

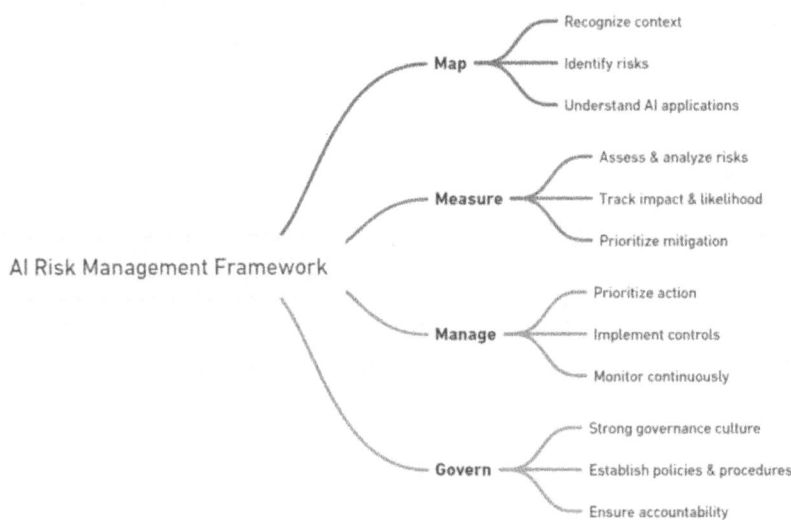

CHAPTER 23   STRATEGIC AI RISK MANAGEMENT & QUANTIFICATION

1. Map

The first step in AI risk management is to recognize the context and identify potential risks related to that context. This involves understanding the specific applications of AI within the organization and the unique risks they pose.

2. Measure

Once risks are identified, they need to be assessed, analyzed, and tracked. This step involves evaluating the potential impact and likelihood of each risk, allowing organizations to prioritize their mitigation efforts.

3. Manage

Managing risks involves prioritizing and acting upon them based on their projected impact. This requires implementing appropriate controls and continuously monitoring the effectiveness of risk mitigation strategies.

4. Govern

At the heart of effective AI risk management is governance. A strong governance framework ensures that a culture of risk management is cultivated and maintained. This involves establishing policies, procedures, and accountability mechanisms to oversee AI risk management activities.

CHAPTER 23   STRATEGIC AI RISK MANAGEMENT & QUANTIFICATION

Key Risk Categories and Mitigation Strategies

AI risk management encompasses several categories of risks, each requiring specific mitigation strategies. The following table summarizes the key risk categories, associated risks, and suggested mitigation approaches.

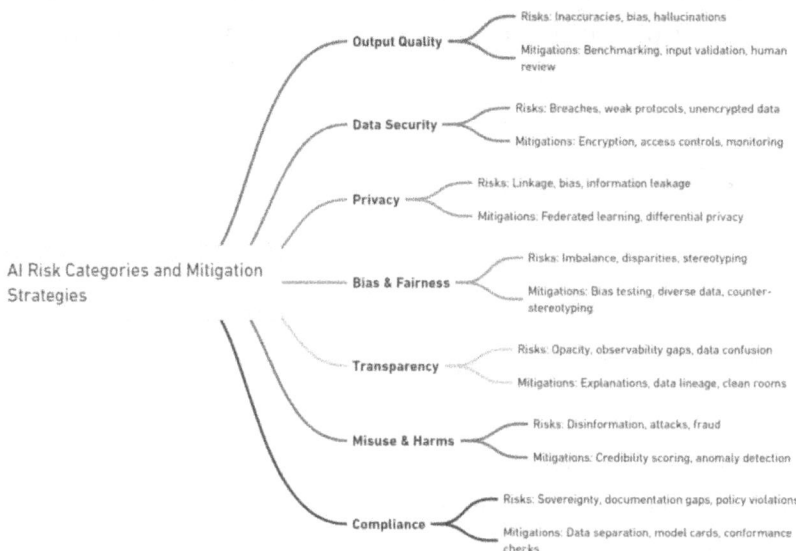

1. Output Quality

- **Risks**: Factual inaccuracies, bias, hallucinations
- **Mitigations**: Accuracy benchmarking, input validation, human-in-the-loop reviews

Ensuring the quality of AI outputs is paramount. This involves implementing measures to validate the accuracy and reliability of AI-generated information, such as benchmarking against known standards and incorporating human oversight.

2. Data Security

- **Risks:** Breaches and theft, weak protocols, unencrypted data, insufficient auditing
- **Mitigations:** Access controls, encryption, activity monitoring

Protecting data is critical for AI systems. Organizations must implement robust security protocols, including encryption and access controls, to safeguard sensitive information from unauthorized access and breaches.

3. Privacy

- **Risks:** Attribute linkage, dataset bias, information leakage
- **Mitigations:** Federated learning, differential privacy, scrubbing tools

Maintaining privacy involves preventing unauthorized access to personal data and ensuring compliance with data protection regulations. Techniques such as federated learning and differential privacy can help protect individual privacy while enabling AI to learn from data.

4. Bias and Fairness

- **Risks:** Representation imbalance, outcome disparities, historical stereotyping
- **Mitigations:** Bias testing, synthetic oversampling, counter stereotypical data

Addressing bias and fairness in AI systems is essential for ethical AI deployment. This involves testing for biases, using diverse training data, and employing techniques to mitigate any identified biases.

5. Transparency

- **Risks**: Black box opacity, observability gaps, data lineage confusion
- **Mitigations**: Local model explanations, metadata persistence, clean room techniques

Ensuring transparency in AI systems helps build trust and allows for better understanding of AI decision-making processes. Techniques such as local model explanations and maintaining data lineage can enhance transparency.

6. Misuse and Harms

- **Risks**: Disinformation, adversarial attacks, fraud
- **Mitigations**: Credibility scoring, anomaly detection, context watermarking

Preventing misuse of AI involves detecting and mitigating potential harms such as disinformation and fraud. Implementing credibility scoring and anomaly detection can help identify and address these issues.

7. Compliance

- **Risks**: Data sovereignty confusion, incomplete documentation, policy violations
- **Mitigations**: Multicloud data separation, model cards, automated conformance checks

CHAPTER 23   STRATEGIC AI RISK MANAGEMENT & QUANTIFICATION

Compliance with legal and regulatory requirements is critical for AI systems. Organizations must ensure that AI deployments adhere to relevant policies and standards through techniques such as automated conformance checks and thorough documentation.

Strategic Framework for Risk Quantification for AI systems

AI risks can be categorized into several levels, each requiring distinct management strategies. The accompanying diagram highlights these levels, illustrating the range from minimal to unacceptable risks. Let's explore these categories and their implications for your organization.

Minimal Risk AI Systems

- **Description**: Includes common AI applications like spam filters and recommendation systems.
- **Management**: Minimal regulation is required. Focus on maintaining baseline security and functionality.

Transparency Risk AI Systems

- **Description**: Encompasses systems at risk of impersonation, manipulation, or deception, such as chatbots and AI-generated content.

- **Management**: Implement strict information and transparency obligations. Ensure that users are aware when they are interacting with AI and not a human.

High-Risk AI Systems

- **Description**: Systems impacting health, safety, or fundamental rights. Examples include AI in healthcare diagnostics and autonomous vehicles.

- **Management**: Conduct thorough conformity assessments, including post-market monitoring. Compliance with rigorous safety standards is essential.

Unacceptable Risk AI Systems

- **Description**: AI applications that violate fundamental rights and values.

- **Management**: These systems are prohibited. Ensure compliance with legal frameworks to avoid severe repercussions.

CHAPTER 23   STRATEGIC AI RISK MANAGEMENT & QUANTIFICATION

Implementing Effective AI Risk Management

To effectively manage AI risks, organizations should follow these strategic recommendations:

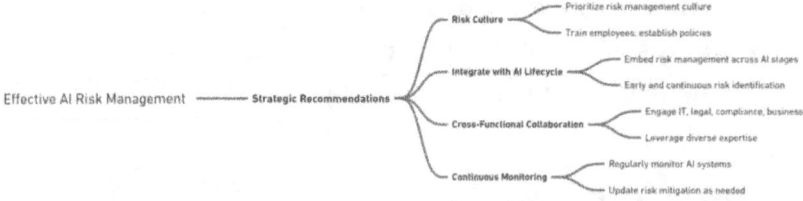

1. Establish a Risk Management Culture

Cultivating a culture that prioritizes risk management is essential. This involves training employees, establishing clear policies, and promoting awareness of AI risks and mitigation strategies.

2. Integrate Risk Management into AI Lifecycle

Risk management should be integrated into every stage of the AI lifecycle, from development to deployment and monitoring. This ensures that risks are identified and addressed early and continuously.

3. Foster Cross-Functional Collaboration

AI risk management requires collaboration across various functions, including IT, legal, compliance, and business units. Cross-functional teams can provide diverse perspectives and expertise to address complex AI risks.

4. Continuous Monitoring and Improvement

AI risk management is an ongoing process. Organizations must continuously monitor AI systems, assess new risks, and update mitigation strategies as needed. This ensures that AI systems remain secure, reliable, and compliant.

Summary

Effective AI risk management is crucial for ensuring the safe and ethical deployment of AI systems. By following a structured framework that includes mapping, measuring, and managing risks, and embedding strong governance practices, organizations can mitigate potential risks and build trust in their AI initiatives. This chapter provides a comprehensive guide to AI risk management, offering strategic insights for corporate boards and senior executives to navigate the complexities of AI deployment and achieve sustainable success.

CHAPTER 24

Leveraging Generative AI: Strategies, Implementation, and Impact

This chapter explores the strategic approaches to implementing Generative AI, focusing on identifying high-impact use cases, optimizing deployment, managing costs, and ensuring sustainability. By following a structured approach, organizations can harness the full potential of Generative AI while navigating the complexities and challenges associated with its deployment.

CHAPTER 24   LEVERAGING GENERATIVE AI: STRATEGIES, IMPLEMENTATION, AND IMPACT

Creating Your Value Hypothesis

The first step in implementing Generative AI is to create a value hypothesis. This strategic assessment evaluates both the potential business value and the feasibility of various Generative AI applications. It should be informed by the organization's purpose, values, current operations, and the competitive and regulatory landscape. A well-defined value hypothesis serves as a benchmark to guide decision-making and assess progress.

1. Strategic Assessment

Begin by conducting a comprehensive analysis of your organization's current state, including financial health, operational efficiency, and market position. Consider the broader business ecosystem and competitive dynamics to identify areas where Generative AI can add value.

2. Benchmarking Potential Value

Compare your value hypothesis with industry benchmarks to estimate the potential long-term value uplift. This helps set realistic expectations and provides a framework for measuring success.

3. Short-Term vs. Long-Term Value

Focus initially on short-term value, such as enhancing productivity and efficiency. However, keep an eye on long-term transformative opportunities that require more time and resources but offer significant strategic benefits.

Prioritizing Key Use Cases

Identifying and prioritizing high-impact use cases is crucial for demonstrating Generative AI's value and achieving buy-in from stakeholders. The top use cases in any industry can generate a substantial portion of the overall value derived from Generative AI.

1. Identifying High-Impact Use Cases

Conduct a thorough analysis to identify Generative AI use cases with the strongest potential to deliver maximum benefit. Focus on areas where Generative AI can significantly enhance performance, reduce costs, or create new revenue streams.

2. Industry-Specific Applications

Different industries will have unique high-impact use cases. For example, in the luxury sector, Generative AI can deliver hyper-personalized marketing, while in software development, Generative AI coding assistants can boost productivity.

3. Assessing Value and Feasibility

Evaluate the potential value of each use case by considering factors such as implementation difficulty, expected return on investment, and alignment with strategic goals. Focus on use cases that offer the highest impact with manageable risks.

Scaling Through Patterns

To maximize value, Generative AI solutions need refinement and scaling. Identifying and leveraging patterns – reusable solutions across different applications – can significantly enhance scalability and efficiency.

1. Model Refinement

Start with refining Generative AI models using your organization's data and additional guardrails to enhance focus and security. This ensures the models deliver relevant and impactful results.

2. Leveraging Patterns

Once initial models are refined, they can often be repurposed for similar uses across the organization. Patterns help identify areas where small additional efforts can adapt existing deployments for broader applications.

3. Value from Net-New Creation and Augmentation

Net-new creation and augmentation offer significant potential value but may require more upfront investment. Focus on developing foundational models that can be adapted for various use cases to drive cumulative value creation.

CHAPTER 24   LEVERAGING GENERATIVE AI: STRATEGIES, IMPLEMENTATION, AND IMPACT

Selecting Foundational Generative AI Tools

Choosing the right foundational Generative AI tools is critical for long-term success. Evaluate technologies, platforms, and service providers to ensure robustness, scalability, and adaptability.

1. Evaluating Technologies

Assess public general access Generative AI models like OpenAI's ChatGPT, Google's Gemini, or Anthropic's Claude for initial use cases. For more sensitive applications, consider secure, private versions of these models.

2. Customizing Models

For specialized uses, such as drafting legal documents or providing personalized services, customize foundational models through retrieval-augmented generation (RAG) or fine-tuning. In some cases, training-bespoke models may be necessary.

3. Avoiding Tech Debt

Anticipate and avoid potential tech debt by selecting technologies that balance robustness with adoptability and adaptability. Ensure that foundational tools can scale and adapt to meet long-term business needs.

Defining Solutions to Maximize Value

Maximize the value of Generative AI by identifying and developing specific solutions that leverage foundational tools and proprietary data.

1. Proprietary Data Integration

Integrate proprietary data to enhance the performance and relevance of Generative AI solutions. This data-driven approach can provide a competitive edge and enable personalized applications.

2. Lateral Thinking and Patterns

Think laterally to identify patterns that can extend the value of initial solutions. For example, a Generative AI chatbot developed for customer service can be adapted for internal assistance, sales support, and training.

3. Incremental Solutions

Develop incremental solutions that build on existing deployments. While individually small, these solutions collectively contribute to substantial value creation and operational efficiency.

Assessing Costs and Carbon Impact

Consider the broad costs of Generative AI implementation, including financial, environmental, and reputational factors. Aim to balance short-term efficiencies with long-term sustainability.

1. Comprehensive Cost Assessment

Evaluate the full spectrum of costs, including development, deployment, and operational expenses. Consider both immediate productivity gains and long-term transformation opportunities.

2. Environmental Impact

Analyze the carbon footprint of Generative AI deployments. While Generative AI can be energy-intensive, efficiency gains can offset emissions. Implement sustainable practices to minimize environmental impact.

3. Reputational Considerations

Consider the potential reputational risks of replacing human labor with Generative AI. Ensure that Generative AI applications are used ethically and responsibly to maintain stakeholder trust and support.

Developing, Testing, and Learning

A continuous cycle of development, testing, and learning is essential for maximizing Generative AI's potential. This iterative process helps refine solutions and adapt to evolving needs.

1. Controlled Deployments

Roll out Generative AI solutions with defined controls and success metrics. Monitor performance and gather insights to inform future enhancements.

2. Iterative Learning

Treat each deployment as an opportunity to learn. Use insights to refine models, enhance capabilities, and adapt strategies.

3. Reevaluating Risks and Governance

Regularly reassess risks and governance frameworks. Ensure that Generative AI deployments remain aligned with organizational goals and comply with ethical standards.

Scaling and Adaptation

Leverage accumulated knowledge and experience to scale Generative AI solutions and adapt them for broader uses. This approach accelerates value creation and enhances organizational agility.

1. Adaptive Scaling

Customize initial solutions for additional uses. Proof of concept and lessons learned from early deployments facilitate faster and more efficient scaling.

2. Broadening Applications

Apply Generative AI patterns to new areas, such as predictive maintenance, logistics management, and precision agriculture. This holistic approach enhances end-to-end operational efficiency.

3. Institutional Knowledge

Build institutional knowledge to support Generative AI adoption and integration. Training and adoption become easier with accumulated expertise and established best practices.

CHAPTER 24   LEVERAGING GENERATIVE AI: STRATEGIES, IMPLEMENTATION, AND IMPACT

Seizing the Generative AI Opportunity

The structured approach outlined in this chapter helps organizations identify, implement, and scale Generative AI solutions effectively. By leveraging the flywheel framework, businesses can establish a virtuous cycle of continuous learning and value creation. This strategic approach enables organizations to harness the transformative power of Generative AI, driving innovation, efficiency, and competitive advantage in an increasingly digital world.

Checklist for Leveraging Generative AI: Strategies, Implementation, and Impact

This checklist is designed for board members and C-suite executives to evaluate and guide the successful implementation of Generative AI strategies. Each question is scored to determine the readiness and thoroughness of the organization in leveraging Generative AI.

Scoring System

- **0 points:** Not addressed
- **1 point:** Partially addressed
- **2 points:** Fully addressed, but needs improvement
- **3 points:** Fully addressed and well-executed

CHAPTER 24   LEVERAGING GENERATIVE AI: STRATEGIES, IMPLEMENTATION, AND IMPACT

Creating Your Value Hypothesis

1. **Strategic Assessment** (Score: 0-3)

   - Has a comprehensive analysis of the organization's current state, including financial health, operational efficiency, and market position, been conducted?

   - Are broader business ecosystem and competitive dynamics considered to identify where Generative AI can add value?

2. **Benchmarking Potential Value** (Score: 0-3)

   - Is the value hypothesis compared with industry benchmarks to estimate potential long-term value uplift?

   - Does the organization have a framework for measuring success against these benchmarks?

3. **Short-Term vs. Long-Term Value** (Score: 0-3)

   - Is there a focus on short-term value enhancement, such as productivity and efficiency?

   - Are long-term transformative opportunities, which offer significant strategic benefits, identified and planned?

Prioritizing Key Use Cases

4. **Identifying High-Impact Use Cases** (Score: 0-3)

   - Are thorough analyses conducted to identify Generative AI use cases with the strongest potential to deliver maximum benefit?

CHAPTER 24   LEVERAGING GENERATIVE AI: STRATEGIES, IMPLEMENTATION, AND IMPACT

- Are focus areas identified where Generative AI can significantly enhance performance, reduce costs, or create new revenue streams?

5. **Industry-Specific Applications** (Score: 0-3)

    - Are high-impact use cases identified for different industries specific to the organization's sector?

    - Is there a clear understanding of how Generative AI can deliver value in unique industry contexts?

6. **Assessing Value and Feasibility** (Score: 0-3)

    - Is the potential value of each use case evaluated considering implementation difficulty, expected ROI, and alignment with strategic goals?

    - Are high-impact, manageable-risk use cases prioritized?

Scaling Through Patterns

7. **Model Refinement** (Score: 0-3)

    - Are Generative AI models refined using the organization's data and additional guardrails to enhance focus and security?

    - Do the models deliver relevant and impactful results after refinement?

8. **Leveraging Patterns** (Score: 0-3)

    - Are initial models repurposed for similar uses across the organization to enhance scalability and efficiency?

- Are patterns identified for small additional efforts that adapt existing deployments for broader applications?

9. **Value from Net-New Creation and Augmentation** (Score: 0-3)

    - Is there a focus on developing foundational models that can be adapted for various use cases?
    - Are potential values from net-new creation and augmentation balanced with the required upfront investment?

Selecting Foundational Generative AI Tools

10. **Evaluating Technologies** (Score: 0-3)

    - Are public general access Generative AI models assessed for initial use cases?
    - For sensitive applications, are secure, private versions of these models considered?

11. **Customizing Models** (Score: 0-3)

    - Are foundational models customized through retrieval-augmented generation (RAG) or fine-tuning for specialized uses?
    - Is there a consideration for training bespoke models if necessary?

12. **Avoiding Tech Debt** (Score: 0-3)

    - Are potential tech debts anticipated and avoided by selecting technologies that balance robustness with adoptability?

    - Can foundational tools scale and adapt to meet long-term business needs?

Defining Solutions to Maximize Value

13. **Proprietary Data Integration** (Score: 0-3)

    - Is proprietary data integrated to enhance the performance and relevance of Generative AI solutions?

    - Does this data-driven approach provide a competitive edge and enable personalized applications?

14. **Lateral Thinking and Patterns** (Score: 0-3)

    - Are patterns identified that extend the value of initial solutions through lateral thinking?

    - Are solutions like Generative AI chatbots adapted for various functions such as internal assistance, sales support, and training?

15. **Incremental Solutions** (Score: 0-3)

    - Are incremental solutions developed that build on existing deployments?

    - Do these small solutions collectively contribute to substantial value creation and operational efficiency?

CHAPTER 24   LEVERAGING GENERATIVE AI: STRATEGIES, IMPLEMENTATION, AND IMPACT

Assessing Costs and Carbon Impact

16. **Comprehensive Cost Assessment** (Score: 0-3)
    - Is the full spectrum of costs evaluated, including development, deployment, and operational expenses?
    - Are both immediate productivity gains and long-term transformation opportunities considered?

17. **Environmental Impact** (Score: 0-3)
    - Is the carbon footprint of Generative AI deployments analyzed?
    - Are sustainable practices implemented to minimize environmental impact?

18. **Reputational Considerations** (Score: 0-3)
    - Are potential reputational risks of replacing human labor with Generative AI considered?
    - Are Generative AI applications used ethically and responsibly to maintain stakeholder trust and support?

Developing, Testing, and Learning

19. **Controlled Deployments** (Score: 0-3)
    - Are Generative AI solutions rolled out with defined controls and success metrics?
    - Is performance monitored and insights gathered to inform future enhancements?

20. **Iterative Learning** (Score: 0-3)

    - Is each deployment treated as an opportunity to learn?

    - Are insights used to refine models, enhance capabilities, and adapt strategies?

21. **Reevaluating Risks and Governance** (Score: 0-3)

    - Are risks and governance frameworks regularly reassessed?

    - Do Generative AI deployments remain aligned with organizational goals and comply with ethical standards?

Scaling and Adaptation

22. **Adaptive Scaling** (Score: 0-3)

    - Are initial solutions customized for additional uses?

    - Do proof of concept and lessons learned from early deployments facilitate faster and more efficient scaling?

23. **Broadening Applications** (Score: 0-3)

    - Are Generative AI patterns applied to new areas, such as predictive maintenance, logistics management, and precision agriculture?

    - Does this holistic approach enhance end-to-end operational efficiency?

24. **Institutional Knowledge** (Score: 0-3)

- Is institutional knowledge built to support Generative AI adoption and integration?
- Are training and adoption processes easier with accumulated expertise and established best practices?

Total Scoring

Each question is scored from 0 to 3, with a maximum possible score of 72 across all 24 questions.

Scoring and Interpretation

- **0–24:** High risk – Immediate action required to address significant gaps in Generative AI strategies.
- **25–48:** Moderate risk – Areas for improvement identified, and a structured plan is needed to enhance Generative AI approaches.
- **49–72:** Low risk – Robust Generative AI strategies in place, but continuous monitoring and minor adjustments are recommended.

Threshold for Passing

Organizations should aim for a minimum score of **48** to ensure they have adequately addressed key Generative AI considerations and are prepared to manage and scale AI operations effectively.

CHAPTER 24   LEVERAGING GENERATIVE AI: STRATEGIES, IMPLEMENTATION, AND IMPACT

Summary

In the high-stakes arena of Generative AI, a well-crafted strategy isn't just a roadmap – it's the difference between unleashing transformative potential and stumbling into a technological quagmire. From crafting a robust value hypothesis to scaling through reusable patterns, the journey of Generative AI implementation demands a delicate balance of visionary thinking and pragmatic execution. As organizations navigate this AI-powered landscape, those who master the art of identifying high-impact use cases, selecting foundational tools wisely, and cultivating a culture of continuous learning will find themselves not just participants in the AI revolution, but architects of industry disruption. In this brave new world, where Generative AI is rewriting the rules of business, will your strategy be the catalyst that propels you to market leadership, or the anchor that leaves you trailing in a sea of missed opportunities?

CHAPTER 25

Evaluating Generative AI Use Cases: A Comprehensive Framework

This chapter outlines a comprehensive framework for assessing Generative AI applications, considering financial impact, disruption potential, and ease of adoption. By leveraging this framework, business leaders can make informed decisions that maximize value and drive strategic success.

Key Considerations in Assessing Generative AI Use Cases

To understand the true potential value of Generative AI use cases, it's essential to look beyond financial projections and consider the likely level of disruption and relative ease of adoption.

This section breaks down the key considerations into several categories, providing a holistic view of the factors that influence the success of Generative AI initiatives.

CHAPTER 25   EVALUATING GENERATIVE AI USE CASES: A COMPREHENSIVE FRAMEWORK

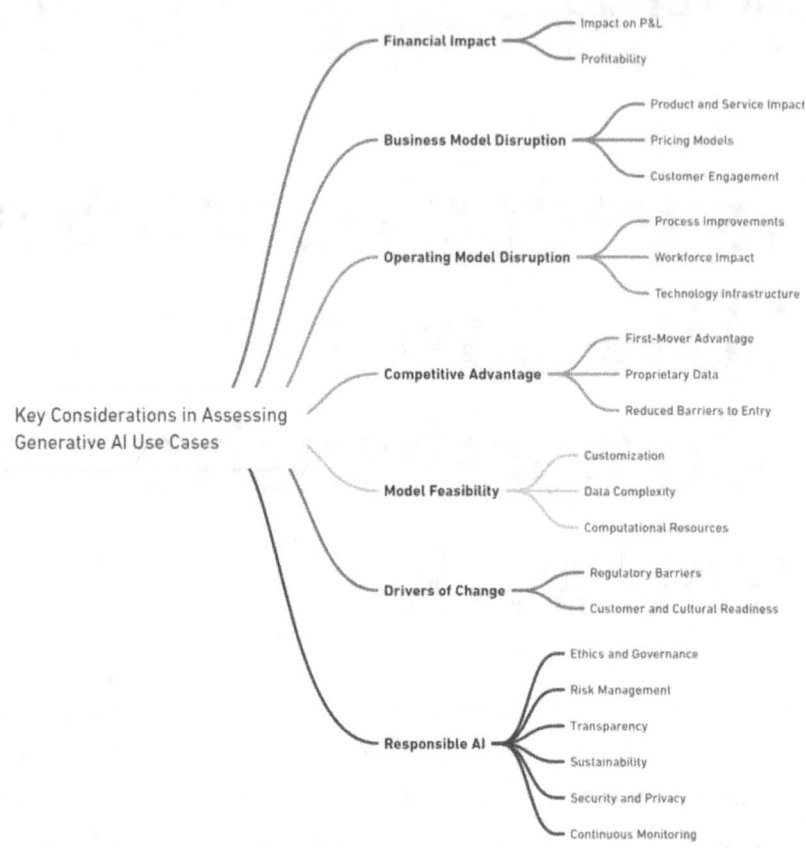

1. Margin (Revenue and Cost)

Evaluating the financial impact of Generative AI use cases involves analyzing their effect on key profit and loss (P&L) line items. Key considerations include

- **Level of Impact on P&L**: Assess how a Generative AI use case can affect revenue streams and cost structures. This includes evaluating the potential for increased sales, cost savings, and overall profitability.

2. Business Model Disruption

Generative AI can significantly alter business models by changing the nature of products and services, pricing strategies, and customer engagement methods. Key considerations include

- **Impact on Products and Services**: Determine how Generative AI will affect your offerings, potentially enabling new products or enhancing existing ones.

- **Changes in Pricing Models**: Evaluate the need to adapt pricing strategies in response to new Generative AI capabilities.

- **Shifts in Customer Engagement**: Analyze how Generative AI can enhance customer interactions and drive loyalty.

3. Operating Model Disruption

Generative AI often necessitates changes in business processes, impacting people, skills, and technology systems. Key considerations include

- **Process Improvements**: Identify opportunities for streamlining and automating processes to enhance efficiency.

- **Impact on Workforce**: Assess the need for reskilling or upskilling employees to work alongside Generative AI.

- **Transformation of Systems**: Ensure that your technology infrastructure can support new Generative AI applications and data requirements.

4. Competitive Disruption

Generative AI offers a competitive edge by providing first-mover advantages, leveraging proprietary data, and reducing barriers to entry. Key considerations include

- **First-Mover Advantage**: Evaluate the benefits of being an early adopter of Generative AI in your industry.

- **Proprietary Data Advantage**: Leverage unique data assets to create differentiated Generative AI applications.

- **Reduced Barriers to Entry**: Understand how Generative AI can lower entry barriers for new competitors and adjust your strategy accordingly.

5. Model Feasibility

The feasibility of deploying Generative AI models depends on the level of customization required, the volume and complexity of training data, and computational demands. Key considerations include

- **Customization Requirements**: Assess the extent to which Generative AI models need to be tailored to specific use cases.

- **Data Volume and Complexity**: Ensure that your data infrastructure can handle the required data for training Generative AI models.

- **Computational Resources**: Evaluate the computational power needed to train and deploy Generative AI models effectively.

6. Drivers of Change

Successful Generative AI implementation requires readiness across regulatory, customer, and cultural dimensions. Key considerations include

- **Regulatory Barriers**: Navigate legal and regulatory challenges associated with Generative AI deployment.
- **Customer Readiness**: Gauge customer acceptance and readiness to engage with Generative AI-enabled products and services.
- **Cultural Readiness**: Foster a culture that embraces innovation and supports Generative AI initiatives.

7. Responsible AI

Ensuring responsible AI deployment involves addressing ethical, governance, and compliance issues. Key considerations include

- **Data and AI Ethics**: Implement ethical guidelines for data usage and AI development.
- **Governance and Compliance**: Establish robust governance frameworks to oversee AI deployment.
- **Risk Management**: Identify and mitigate risks associated with Generative AI, including bias and fairness.
- **Transparency and Explainability**: Ensure that Generative AI models are transparent and their outputs explainable.
- **Sustainability**: Consider the environmental impact of Generative AI and strive for sustainable practices.

- **Security and Privacy**: Protect data and ensure privacy through robust security measures.
- **Validation and Monitoring**: Continuously validate and monitor Generative AI models to maintain their reliability and performance.

Case Study in Generative AI Deployment: A Global Beverage Company

To illustrate the practical application of this framework, consider a global beverage company leveraging Generative AI to optimize its operations:

1. Initial Focus on Predictive Maintenance

The company began by using Generative AI for predictive maintenance in its factories. The AI tool identified patterns that could lead to equipment failures, allowing for proactive maintenance and reducing downtime.

2. Scaling to Logistics Management

Building on the success of predictive maintenance, the company adapted the same Generative AI patterns to enhance transport and logistics management. This extension provided valuable insights into optimizing supply chains and improving delivery efficiency.

3. Expanding to Precision Agriculture

Further leveraging Generative AI, the company applied the technology to precision agriculture, enhancing crop production efficiency. This holistic approach enabled an end-to-end view of the supply chain, integrating predictive analytics to drive system-wide improvements.

4. Continuous Learning and Adaptation

Throughout the deployment, the company emphasized continuous learning. Each Generative AI application was tested, refined, and adapted based on performance metrics and stakeholder feedback. This iterative process ensured that the AI tools remained relevant and impactful.

Checklist for Evaluating Generative AI Use Cases

Scoring System

- **0 points:** Not addressed
- **1 point:** Partially addressed
- **2 points:** Fully addressed, but needs improvement
- **3 points:** Fully addressed and well-executed

Organizations should aim for a minimum score of **45** out of a possible 84 to ensure they have adequately considered all critical aspects of Generative AI use cases.

Financial Impact and Business Model Disruption

1. **Level of Impact on P&L**
    - Does the Generative AI use case have a clear and significant impact on revenue streams? (Score: 0-3)
    - Is there a potential for cost savings and improved profitability through the use of Generative AI? (Score: 0-3)

2. **Impact on Products and Services**
   - Can Generative AI enable the development of new products or enhance existing ones? (Score: 0-3)
   - Are there changes required in pricing models to accommodate Generative AI capabilities? (Score: 0-3)

3. **Shifts in Customer Engagement**
   - Will Generative AI improve customer interactions and drive loyalty? (Score: 0-3)
   - Does the use case align with the strategic objectives of enhancing customer engagement? (Score: 0-3)

Operating Model and Competitive Disruption

4. **Process Improvements**
   - Are there opportunities to streamline and automate processes using Generative AI? (Score: 0-3)
   - Will the AI implementation significantly enhance operational efficiency? (Score: 0-3)

5. **Impact on Workforce**
   - Is there a need for reskilling or upskilling employees to work alongside Generative AI? (Score: 0-3)
   - Does the organization have plans to manage workforce changes effectively? (Score: 0-3)

6. **Transformation of Systems**
   - Can the existing technology infrastructure support new Generative AI applications? (Score: 0-3)
   - Are there plans to upgrade or adapt systems to integrate AI solutions? (Score: 0-3)

7. **First-Mover Advantage**
   - Is the organization poised to gain a first-mover advantage with this Generative AI use case? (Score: 0-3)
   - Does the use case provide a significant competitive edge? (Score: 0-3)

Feasibility and Readiness

8. **Customization Requirements**
   - To what extent do Generative AI models need to be tailored to specific use cases? (Score: 0-3)
   - Are there resources available for necessary customization? (Score: 0-3)

9. **Data Volume and Complexity**
   - Does the organization's data infrastructure support the volume and complexity of data required? (Score: 0-3)
   - Are data management practices in place to handle this data effectively? (Score: 0-3)

10. **Computational Resources**
    - Are sufficient computational resources available to train and deploy Generative AI models? (Score: 0-3)
    - Does the organization have a plan to manage these resources efficiently? (Score: 0-3)

11. **Regulatory Barriers**
    - Are there legal and regulatory challenges associated with deploying Generative AI? (Score: 0-3)
    - Does the organization have strategies to navigate these barriers? (Score: 0-3)

12. **Customer Readiness**
    - Are customers prepared to engage with Generative AI-enabled products and services? (Score: 0-3)
    - Has there been any feedback or pilot testing to gauge customer readiness? (Score: 0-3)

13. **Cultural Readiness**
    - Does the organization's culture support innovation and Generative AI initiatives? (Score: 0-3)
    - Are there measures in place to foster an AI-friendly culture? (Score: 0-3)

Responsible AI and Risk Management

14. **Data and AI Ethics**
    - Are ethical guidelines for data usage and AI development in place? (Score: 0-3)
    - Does the organization actively promote ethical AI practices? (Score: 0-3)

15. **Governance and Compliance**
    - Is there a robust governance framework to oversee AI deployment? (Score: 0-3)
    - Are compliance measures adhered to consistently? (Score: 0-3)

16. **Risk Management**
    - Have risks associated with Generative AI, such as bias and fairness, been identified? (Score: 0-3)
    - Are there plans to mitigate these risks effectively? (Score: 0-3)

17. **Transparency and Explainability**
    - Are Generative AI models transparent and their outputs explainable? (Score: 0-3)
    - Is there an effort to ensure that AI decisions can be understood and trusted? (Score: 0-3)

18. **Sustainability**
    - Has the environmental impact of Generative AI been considered? (Score: 0-3)
    - Are sustainable practices being implemented to minimize the carbon footprint? (Score: 0-3)

19. **Security and Privacy**
    - Are robust security measures in place to protect data and ensure privacy? (Score: 0-3)
    - Is there continuous monitoring to safeguard against breaches? (Score: 0-3)

20. **Validation and Monitoring**
    - Are Generative AI models continuously validated and monitored? (Score: 0-3)
    - Does the organization have a system in place for regular model updates and performance checks? (Score: 0-3)

Total Scoring

Each question is scored from 0 to 3, with a maximum possible score of 60 across all 20 questions.

Scoring and Interpretation

- **0–20:** High risk – Immediate action required to address significant gaps in evaluating Generative AI use cases.
- **21–40:** Moderate risk – Areas for improvement identified, and a structured plan is needed to enhance AI use case evaluation.
- **41–60:** Low risk – Robust AI evaluation strategies in place, but continuous monitoring and minor adjustments are recommended.

Threshold for Passing

Organizations should aim for a minimum score of **45** to ensure they have adequately considered all critical aspects of evaluating Generative AI use cases and are prepared to manage and scale AI operations effectively

Summary

By following the structured approach outlined in this chapter, organizations can effectively evaluate and prioritize Generative AI use cases. This comprehensive framework helps identify high-impact applications, optimize deployment strategies, and ensure responsible AI use. Embracing this approach allows businesses to harness the transformative power of Generative AI, driving innovation, efficiency, and competitive advantage in an ever-evolving digital landscape.

CHAPTER 26

AI Executive Compensation: Insights from Europe and the United States

Organizational Structure and Roles

AI leadership roles are diverse, reflecting the multifaceted nature of AI integration within businesses. The survey highlights several key roles that AI leaders occupy, including Chief Data and Analytics Officer, Chief Data Scientist, and Head of Machine Learning or Artificial Intelligence.

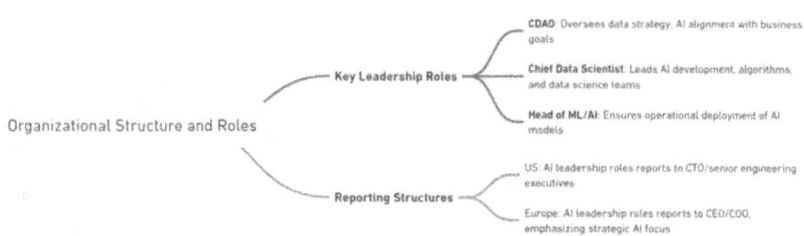

CHAPTER 26 AI EXECUTIVE COMPENSATION: INSIGHTS FROM EUROPE AND THE UNITED STATES

1. Key Leadership Roles

- **Chief Data and Analytics Officer (CDAO)**: This role often involves overseeing the entire data strategy, including data governance, analytics, and AI initiatives. CDAOs are pivotal in aligning AI projects with business goals.
- **Chief Data Scientist**: Focused on the technical aspects of AI, Chief Data Scientists lead the development of algorithms, machine learning models, and data science teams.
- **Head of Machine Learning or Artificial Intelligence**: This role emphasizes the operational deployment of AI technologies, ensuring that AI models are effectively integrated into business processes.

2. Experience and Background

AI leaders typically have extensive experience in sectors like financial services, technology, and consumer industries. The survey indicates that financial services is the most common sector for AI leaders in both Europe and the United States, followed by consumer, retail, and media industries.

3. Reporting Structures

The reporting structure for AI leaders varies significantly between regions. In the United States, AI leaders often report to the Chief Technology Officer (CTO) or senior engineering executives, reflecting a more established integration of AI within the technology function. In Europe, AI leaders are more likely to report directly to the CEO or Chief Operating Officer (COO), indicating a strategic emphasis on AI within the business leadership.

CHAPTER 26   AI EXECUTIVE COMPENSATION: INSIGHTS FROM EUROPE AND THE UNITED STATES

Compensation Insights

Compensation for AI leaders varies widely based on region, industry, and the size of the organization. The survey provides a detailed comparison of base salaries, bonuses, and equity components for AI executives in Europe and the United States.

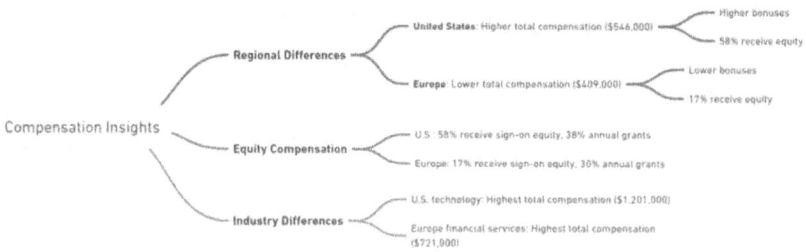

1. Base Salaries and Bonuses

United States: The median base salary for AI leaders in the United States is significantly higher than in Europe, with a median total cash compensation (including bonuses) of $546,000. Bonuses in the United States are also higher, reflecting a greater emphasis on performance-based compensation.

Europe: In Europe, the median total cash compensation is $409,000. While base salaries are competitive, bonuses are generally lower than in the United States. This difference highlights varying compensation philosophies and market conditions between the two regions.

2. Equity Compensation

Equity compensation is a crucial component of total remuneration for AI leaders, particularly in the United States. In the United States, 58% of AI executives receive sign-on equity in the form of restricted stock units (RSUs)

or performance share units (PSUs), compared to 17% in Europe. Annual equity grants are also more common in the United States, with 38% of executives receiving RSUs annually, compared to 30% in Europe.

3. Industry-Specific Compensation

Compensation levels also vary significantly by industry. In the United States, the technology sector offers the highest total compensation, with a median total package of $1,201,000. In Europe, financial services lead with a median total compensation of $721,000. These differences reflect the varying demand for AI talent and the strategic importance of AI across different sectors.

Additional Insights from Compensation Snapshot

1. Financial Services Dominance

In the United States, AI leaders in the financial services sector receive a median total compensation of $1,072,000, significantly higher than other industries. This underscores the value placed on AI leadership in financial services, where data-driven decision-making and risk management are critical.

2. Team Size and Compensation

Compensation also varies with the size of the team managed by AI leaders. Executives leading teams of 51–100 individuals have a higher median total compensation ($1,011,000) compared to those managing smaller teams. This trend is consistent across both regions, highlighting the increased responsibilities and complexities associated with larger teams.

3. Regional Variations within the United States

Within the United States, regional variations exist in AI leadership compensation. For example, AI leaders on the West Coast, a tech hub, receive a median total compensation of $1,013,000, reflecting the high demand for AI talent in this region. In contrast, AI leaders in the Southeast have a lower median total compensation ($791,000), indicating regional disparities in compensation practices.

4. Gender and Ethnic Disparities

The survey also highlights disparities in compensation based on gender and ethnicity. Male AI leaders have a higher median total compensation ($929,000) compared to their female counterparts ($886,000). Similarly, Caucasian executives receive higher compensation compared to other ethnic groups, emphasizing the need for more equitable compensation practices in AI leadership.

Diversity and Inclusion

Diversity remains a challenge in the AI leadership landscape, with significant disparities in gender and ethnicity representation.

1. Gender Diversity

The majority of AI executives are male, particularly in Europe where only 10% of AI leaders are women. In the United States, women represent 20% of AI leaders, indicating slightly better, but still insufficient, gender diversity.

CHAPTER 26 AI EXECUTIVE COMPENSATION: INSIGHTS FROM EUROPE AND THE UNITED STATES

2. Ethnic Diversity

Ethnic diversity also shows room for improvement. In the United States, 45% of AI executives are non-white, with Asian and Asian American professionals being the most represented minority group at 33%. In Europe, non-white executives constitute only 19% of the AI leadership, highlighting a need for more inclusive hiring practices.

Key Considerations for AI Leadership

Organizations must consider several factors when hiring and retaining AI leaders to ensure they attract top talent and foster a conducive environment for AI innovation.

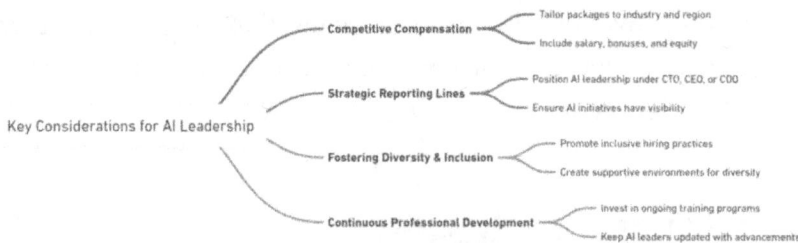

1. Competitive Compensation Packages

Offering competitive compensation packages, including base salary, bonuses, and equity, is essential to attract top AI talent. As the survey indicates, compensation expectations can vary significantly by region and industry, necessitating tailored strategies.

2. Strategic Reporting Lines

Aligning AI leadership roles with strategic reporting lines ensures that AI initiatives receive the necessary support and visibility within the organization. Whether reporting to the CTO, CEO, or COO, the positioning of AI leaders within the organizational hierarchy can significantly impact their effectiveness.

3. Fostering Diversity and Inclusion

Promoting diversity and inclusion within AI leadership not only enhances innovation but also reflects broader societal values. Implementing inclusive hiring practices and creating supportive environments for underrepresented groups are critical steps toward achieving this goal.

4. Continuous Professional Development

Given the rapid evolution of AI technologies, continuous professional development is crucial for AI leaders. Organizations should invest in ongoing training and development programs to keep their AI leaders at the forefront of technological advancements.

Checklist for AI Executive Compensation Insights from Europe and the United States

Scoring System

- **0 points:** Not addressed
- **1 point:** Partially addressed
- **2 points:** Fully addressed, but needs improvement
- **3 points:** Fully addressed and well-executed

CHAPTER 26 AI EXECUTIVE COMPENSATION: INSIGHTS FROM EUROPE AND THE UNITED STATES

Common Themes

1. **Alignment with Strategic Goals** (Score: 0-3)
   - Does the compensation structure for AI leaders align with the organization's strategic goals, particularly in AI integration and innovation?

2. **Market Competitiveness** (Score: 0-3)
   - Is the compensation package competitive within the industry and region to attract and retain top AI talent?

3. **Equity and Fairness** (Score: 0-3)
   - Are compensation packages designed to be fair and equitable across gender, ethnicity, and other diversity factors?

4. **Incentives for Performance and Innovation** (Score: 0-3)
   - Are bonuses and equity components tied to clear performance metrics and innovation targets in AI initiatives?

5. **Diversity and Inclusion** (Score: 0-3)
   - Are there strategies in place to improve gender and ethnic diversity in AI leadership roles?

Specific Compensation Components

Base Salaries and Bonuses

6. **Region-Specific Salary Competitiveness** (Score: 0-3)

    - Does the organization offer base salaries and bonuses that are competitive within the specific region (e.g., United States vs. Europe)?

7. **Performance-Based Bonuses** (Score: 0-3)

    - Are bonuses structured to reward measurable performance outcomes in AI projects?

Equity Compensation

8. **Sign-On Equity Packages** (Score: 0-3)

    - Does the organization offer sign-on equity packages (e.g., RSUs, PSUs) to AI leaders, particularly in the United States?

9. **Annual Equity Grants** (Score: 0-3)

    - Are annual equity grants provided to AI leaders to ensure long-term alignment with company goals?

Industry-Specific Compensation

10. **Sector-Specific Compensation Adjustments** (Score: 0-3)

    - Are compensation packages tailored to reflect the strategic importance of AI in specific sectors (e.g., technology, financial services)?

11. **Team Size Considerations** (Score: 0-3)
    - Does the compensation reflect the size and complexity of the teams managed by AI leaders?

Diversity and Inclusion

12. **Gender Pay Equity** (Score: 0-3)
    - Is there parity in compensation between male and female AI leaders?
13. **Ethnic Diversity in Compensation** (Score: 0-3)
    - Are compensation packages equitable across different ethnic groups, ensuring non-discriminatory practices?

Reporting Structures

14. **Strategic Reporting Lines** (Score: 0-3)
    - Are AI leaders positioned within the organizational structure to maximize their strategic impact (e.g., reporting to CTO, CEO, COO)?

Professional Development and Retention

15. **Investment in Continuous Learning** (Score: 0-3)
    - Does the organization provide opportunities for continuous professional development for AI leaders?

16. **Retention Strategies** (Score: 0-3)

    - Are there effective retention strategies in place, including long-term incentives like equity vesting?

Total Scoring

Each question is scored from 0 to 3, with a maximum possible score of 48 across all 16 questions.

Scoring and Interpretation

- **0–16:** High risk – Immediate action required to address significant gaps in AI executive compensation strategies.

- **17–32:** Moderate risk – Areas for improvement identified, with a structured plan needed to enhance compensation practices.

- **33–48:** Low risk – Robust compensation strategies in place, but continuous monitoring and minor adjustments are recommended.

Threshold for Passing

Organizations should aim for a minimum score of **32** to ensure they have adequately addressed key aspects of AI executive compensation, fostering an environment that attracts and retains top talent while promoting fairness and equity

CHAPTER 26   AI EXECUTIVE COMPENSATION: INSIGHTS FROM EUROPE AND
            THE UNITED STATES

Summary

The insights from the 2021 Data Analytics and AI Executive Organization and Compensation Survey underscore the critical role of AI leaders in driving business transformation. By understanding the organizational structures, compensation trends, and diversity challenges, businesses can better position themselves to attract and retain top AI talent. As AI continues to evolve, the role of AI leaders will become increasingly pivotal in shaping the future of industries worldwide. This chapter provides a roadmap for organizations to navigate the complexities of AI leadership and build a strong foundation for AI-driven success.

CHAPTER 27

Strategic Insights on the Reporting Structures of AI Executives

In the rapidly evolving landscape of artificial intelligence (AI), the organizational positioning of AI executives is paramount to the success of AI initiatives. This chapter explores the strategic implications of AI executives' reporting structures, providing direction for corporate boards and senior executives to optimize their leadership frameworks. By examining broad patterns and offering strategic recommendations, this chapter aims to guide organizations in leveraging AI leadership effectively to drive innovation and maintain a competitive edge.

CHAPTER 27   STRATEGIC INSIGHTS ON THE REPORTING STRUCTURES OF AI EXECUTIVES

Broad Patterns in Reporting Structures

The placement of AI executives within an organization reflects its strategic priorities and influences the effectiveness of AI initiatives. The 2021 Data Analytics and AI Executive Organization and Compensation Survey reveals distinct regional patterns and organizational strategies:

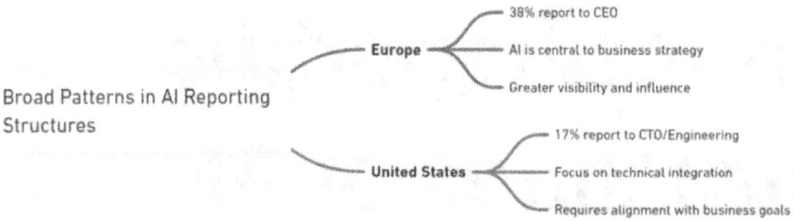

Broad Patterns in AI Reporting Structures

- **High-Level Reporting in Europe**: In Europe, a significant proportion of AI executives (38%) report directly to the CEO. This trend underscores the strategic importance placed on AI, indicating that AI initiatives are central to the overall business strategy. Reporting to the CEO provides AI executives with greater visibility and influence, facilitating the alignment of AI projects with corporate objectives.

- **Technical Emphasis in the United States**: In the United States, AI executives are more likely to report to the CTO or senior engineering executives (17%). This pattern highlights a strong focus on the technical and engineering aspects of AI, ensuring that AI technologies are effectively integrated into the company's IT infrastructure. However, this technical emphasis may require additional efforts to align AI initiatives with broader business goals.

CHAPTER 27 STRATEGIC INSIGHTS ON THE REPORTING STRUCTURES OF AI EXECUTIVES

- **Operational Integration:** Reporting to the COO or Chief Administrative Officer, seen in both regions but more common in Europe (15%), emphasizes the operational integration of AI. This structure focuses on embedding AI into core business processes to enhance efficiency and productivity.

Strategic Directions for Executives and Boards

Understanding these patterns provides a foundation for strategic decision-making regarding AI leadership. Executives and boards should consider the following directions to optimize the impact of their AI initiatives:

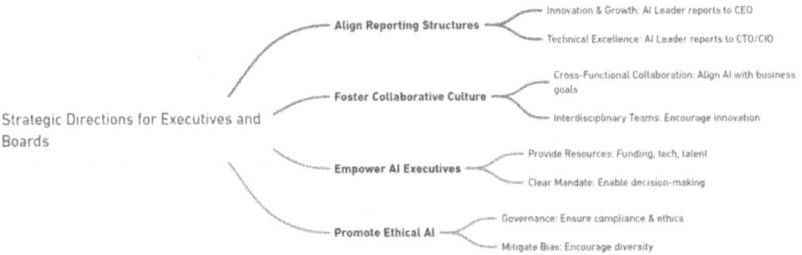

Align Reporting Structures with Strategic Objectives

- **For Innovation and Growth:** If AI is a top strategic priority driving innovation and growth, AI executives should report directly to the CEO. This alignment ensures that AI initiatives receive the necessary support and visibility at the highest organizational level, facilitating strategic alignment and resource allocation.

- **For Technical Excellence**: If the focus is on technical implementation and integration, AI executives should report to the CTO or CIO. This structure supports the development and deployment of robust AI technologies, ensuring technical excellence and operational integrity.

Foster a Collaborative Culture

- **Cross-Functional Collaboration**: AI initiatives often require collaboration across various functions such as marketing, operations, and sales. Boards and executives should promote a culture of collaboration, encouraging AI leaders to engage with other departments to ensure comprehensive integration and alignment with business objectives.
- **Interdisciplinary Teams**: Forming interdisciplinary teams that include members from different business units can enhance the effectiveness of AI initiatives. These teams bring diverse perspectives, fostering innovation and ensuring that AI solutions address a broad range of business challenges.

Empower AI Executives with Resources and Authority

- **Adequate Resourcing**: AI leaders need access to sufficient resources, including funding, technology, and talent. Boards should ensure that AI initiatives are well-resourced to drive successful outcomes.

- **Clear Mandate and Authority**: Providing AI executives with a clear mandate and the authority to make decisions is crucial. This empowerment enables AI leaders to act decisively, implement projects efficiently, and drive strategic initiatives.

Promote Ethical and Responsible AI

- **Governance and Compliance**: Boards should establish robust governance frameworks to oversee AI initiatives, ensuring compliance with ethical standards and regulatory requirements. AI executives should report on these aspects regularly to maintain transparency and accountability.

- **Bias and Fairness**: Promoting diversity within AI leadership and teams can help mitigate biases in AI models. Boards should encourage inclusive hiring practices and foster a culture that values diverse perspectives.

Strategic Recommendations

To maximize the impact of AI leadership, organizations should adopt the following strategic recommendations:

- **Strategic Alignment**: Ensure that the reporting structure of AI executives aligns with the organization's strategic priorities. This alignment enhances the effectiveness of AI initiatives by integrating them into the core business strategy.

- **Resource Allocation**: Provide AI leaders with the necessary resources, including budget, talent, and technology. Adequate resourcing is essential for the successful implementation and scaling of AI projects.

- **Empowerment and Authority**: Empower AI executives with the authority to drive initiatives and make strategic decisions. Clear mandates and decision-making authority enable AI leaders to act swiftly and effectively.

- **Cross-Functional Integration**: Foster a collaborative environment where AI leaders work closely with other business functions. This integration ensures that AI initiatives are aligned with broader business goals and enhance overall organizational performance.

- **Ethical Oversight**: Establish governance structures to ensure that AI initiatives are ethical and compliant with regulatory standards. Boards should monitor AI projects for bias, fairness, and transparency to maintain trust and accountability.

CHAPTER 27 STRATEGIC INSIGHTS ON THE REPORTING STRUCTURES OF AI EXECUTIVES

Checklist for AI Executive Reporting Structures: A Strategic Guide for Boards and C-Suites

Scoring System

- **0 points:** Not addressed
- **1 point:** Partially addressed
- **2 points:** Fully addressed, but needs improvement
- **3 points:** Fully addressed and well-executed

Common Themes for All Reporting Structures

1. **Alignment with Strategic Objectives**
   (Score: 0-3 each)

   - Is the reporting structure of AI executives aligned with the organization's broader strategic goals?
   - Does the reporting structure facilitate the integration of AI initiatives into the core business strategy?

2. **Cross-Functional Collaboration** (Score: 0-3 each)

   - Is there a culture that promotes collaboration between AI executives and other key departments such as marketing, operations, and sales?
   - Are interdisciplinary teams formed to bring diverse perspectives to AI initiatives?

3. **Resourcing and Authority** (Score: 0-3 each)
   - Do AI executives have access to sufficient resources, including funding, technology, and talent?
   - Are AI executives empowered with the authority to make strategic decisions and implement projects efficiently?

4. **Ethical Governance and Compliance** (Score: 0-3 each)
   - Are there robust governance frameworks in place to oversee AI initiatives, ensuring compliance with ethical standards and regulatory requirements?
   - Are AI leaders required to report regularly on issues related to bias, fairness, and transparency?

Specific Reporting Structures
AI Executives Reporting Directly to the CEO

5. **Strategic Influence and Visibility** (Score: 0-3 each)
   - Does the direct reporting to the CEO enhance the visibility and strategic influence of AI initiatives within the organization?
   - Are AI projects receiving the necessary support and alignment at the highest organizational level?

CHAPTER 27 STRATEGIC INSIGHTS ON THE REPORTING STRUCTURES OF AI EXECUTIVES

AI Executives Reporting to the CTO/CIO

6. **Technical Excellence and Integration**
   (Score: 0-3 each)

   - Is the focus on technical implementation and integration of AI technologies effectively managed under the CTO/CIO?

   - Does this structure ensure that AI initiatives are developed and deployed with robust technical standards?

AI Executives Reporting to the COO/Chief Administrative Officer

7. **Operational Integration** (Score: 0-3 each)

   - Does the reporting to the COO/CAO facilitate the seamless integration of AI into core business processes?

   - Are AI initiatives contributing to enhanced operational efficiency and productivity?

Strategic Recommendations

8. **Strategic Alignment and Empowerment**
   (Score: 0-3 each)

   - Is the reporting structure strategically aligned to maximize the impact of AI leadership within the organization?

   - Are AI leaders empowered with clear mandates and the authority to drive initiatives?

9. **Resource Allocation and Support** (Score: 0-3 each)
   - Are AI executives provided with the necessary resources and support to successfully implement and scale AI projects?
   - Is there a system in place to ensure ongoing resource allocation as AI initiatives evolve?

10. **Ethical Oversight and Transparency** (Score: 0-3 each)
    - Are governance structures robust enough to ensure ethical AI practices and transparency in reporting?
    - Does the board monitor AI projects for bias, fairness, and compliance with regulatory standards?

Total Scoring

Each question is scored from 0 to 3, with a maximum possible score of 90 across all 30 questions.

Scoring and Interpretation

- **0–30**: High risk – Immediate action required to address significant gaps in AI executive reporting structures and strategic alignment.
- **31–60**: Moderate risk – Areas for improvement identified; structured plans are needed to optimize AI leadership and its alignment with organizational goals.

- **61–90**: Low risk – Robust reporting structures and strategic alignment in place, but continuous monitoring and adjustments are recommended.

Threshold for Passing

Organizations should aim for a minimum score of **60** to ensure that the AI executive reporting structure is strategically aligned with broader business objectives and effectively supports AI initiatives.

Summary

The strategic positioning of AI executives within an organization is a critical determinant of the success of AI initiatives. By understanding broad patterns in reporting structures and adopting strategic recommendations, corporate boards and senior executives can optimize their AI leadership frameworks. This approach will ensure that AI initiatives are effectively integrated into business strategies, driving innovation, enhancing operational efficiency, and maintaining a competitive advantage in the evolving business landscape.

CHAPTER 28

Governance and Oversight of AI Systems

Effective governance and oversight are essential for maintaining the trustworthiness of AI systems. Boards must ensure that AI governance frameworks are robust and integrate inputs from various business functions.

Key Governance Strategies

- **Establish AI Governance Committees**: Form dedicated committees to oversee AI initiatives, ensuring cross-functional collaboration and comprehensive oversight.

- **Develop AI Policies and Procedures**: Create and enforce policies that guide the ethical and responsible use of AI within the organization.

- **Regular AI Audits**: Conduct regular audits to assess the performance, fairness, and compliance of AI systems.

Practical Steps for Boards

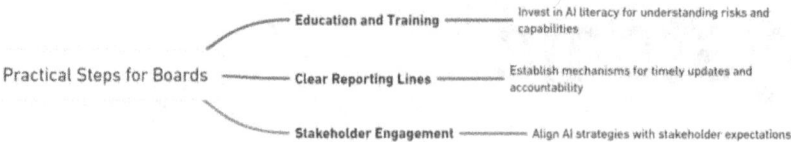

- **Education and Training**: Invest in AI literacy programs for board members to understand AI's capabilities and risks.

- **Clear Reporting Lines**: Establish clear reporting mechanisms for AI initiatives to ensure timely updates and accountability.

- **Stakeholder Engagement**: Engage with stakeholders, including employees, customers, and regulators, to align AI strategies with their expectations and concerns.

Ensuring Ethical AI Practices

Ethical considerations are integral to the development and deployment of trustworthy AI systems. Boards must ensure that AI practices align with the organization's ethical standards and societal values.

CHAPTER 28 GOVERNANCE AND OVERSIGHT OF AI SYSTEMS

Ethical Guidelines for AI

- **Bias Mitigation**: Implement strategies to identify and mitigate biases in AI systems.
- **Transparency**: Ensure transparency in AI operations and decision-making processes.
- **Privacy Protection**: Uphold stringent privacy standards to protect user data.

Monitoring and Evaluation

- **Continuous Monitoring**: Regularly monitor AI systems to detect and address ethical issues.
- **Impact Assessments**: Conduct impact assessments to evaluate the ethical implications of AI applications.
- **Stakeholder Feedback**: Incorporate feedback from stakeholders to continuously improve AI practices.

CHAPTER 28   GOVERNANCE AND OVERSIGHT OF AI SYSTEMS

Understanding AI's Role in Corporate Strategy

AI can transform various facets of business operations and strategy. For boards, understanding AI's potential is key to leveraging its benefits for improved decision-making and operational efficiency.

Strategic Opportunities of AI

- **Enhanced Decision-Making:** AI enables sophisticated data analysis, helping to identify trends that can refine financial projections and guide capital allocation.

- **Risk Management:** AI can foresee potential risks, aiding in proactive risk mitigation and outcome prediction.

- **Operational Efficiency:** AI can streamline information and reporting systems, ensuring timely identification of issues and enhancing the board's decision-making capabilities.

Board's Oversight Role

- **In-depth Understanding:** Boards must thoroughly understand how AI is being utilized within their organization and the associated strategic opportunities and risks, including those related to algorithm and data bias.

- **Ethical Use:** It is essential to ensure that AI applications align with the company's values and ethical standards, fostering responsible use.

CHAPTER 28   GOVERNANCE AND OVERSIGHT OF AI SYSTEMS

Establishing Effective AI Governance

AI governance should include clear reporting structures, regular updates on AI initiatives, and metrics to measure AI's impact. The board should treat AI oversight with the same rigor as any other significant business matter.

Governance Framework

- **Strategic Disruption and Opportunities**: Evaluating how AI could disrupt the industry and what strategic opportunities it presents
- **AI Utilization**: Understanding the role of AI in company processes and third-party products
- **Data Governance**: Ensuring robust governance of data used for AI, including risk management for data handling
- **AI Governance Systems**: Reviewing the AI governance framework, ensuring it integrates insights from all relevant business functions

Evaluating AI Projects

- **Strategic Alignment**: Assessing if AI initiatives align with the company's strategic goals
- **Resource Assessment**: Determining if the company has the necessary expertise and resources to pursue AI responsibly
- **Resilience and Compliance**: Ensuring AI initiatives are resilient in terms of cybersecurity and operational integrity
- **Success Metrics**: Defining clear metrics for measuring the success of AI projects and establishing proof-of-concept protocols

Practical Steps for AI Oversight

Boards should adopt a proactive stance towards AI oversight, leveraging external expertise and staying informed about regulatory changes.

Proactive Oversight Actions

- **AI Education**: Investing in AI literacy for board members to understand AI concepts, capabilities, and limitations

- **Developing AI Strategies**: Crafting clear AI strategies that align with corporate goals, focusing on responsible use and risk management

- **Regular Risk Assessments**: Conducting ongoing risk assessments to identify potential AI-related threats

- **Third-Party Audits**: Engaging third parties to audit AI systems and identify vulnerabilities

- **Crisis Management Planning**: Preparing for potential AI-related issues, such as data breaches or ethical concerns

- **External Collaboration**: Consulting with AI experts, legal advisors, and consultants to stay updated on emerging AI risks and mitigation strategies

Mitigating Directors' and Officers' Liability

- **Compliance and Ethics**: Ensuring AI applications comply with ethical standards and legal regulations

- **Data Protection**: Implementing robust data protection measures to prevent breaches and ensure compliance with privacy laws

- **Documentation of Oversight**: Recording AI oversight activities in board minutes and presentations

Leveraging AI for Business Value

AI can drive significant improvements in efficiency and revenue. Boards should explore integrating AI into various business processes to maximize its potential.

Business Value Applications

- **Due Diligence**: Utilizing AI to enhance due diligence processes by analyzing large datasets and identifying risks

- **Customer Engagement**: Employing AI-driven marketing to improve customer acquisition and retention

- **Compliance Automation**: Automating data collection, analysis, and reporting to minimize compliance risks

- **Innovation and R&D**: Using AI to expedite product development and improve customer experiences

Strategic Considerations for Boards

Boards must stay informed about the evolving regulatory landscape and proactively address AI-related challenges.

Key Board Responsibilities

- **Regulatory Awareness**: Keeping abreast of new regulations and legislative changes related to AI
- **Policy Review**: Regularly reviewing and updating AI-related policies
- **Ethical Oversight**: Discussing the ethical implications of AI with management and ensuring proper risk mitigation strategies
- **Performance Monitoring**: Tracking AI's role in employee performance assessments and ensuring regulatory compliance

Checklist for Governance and Oversight of AI Systems

Scoring System

- 0 points: Not addressed
- 1 point: Partially addressed
- 2 points: Fully addressed, but needs improvement
- 3 points: Fully addressed and well-executed

Governance and Oversight

1. **AI Governance Committees** (Score: 0-3 each)
   - Have we established dedicated AI governance committees?
   - Is there cross-functional collaboration within these committees?

2. **AI Policies and Procedures** (Score: 0-3 each)
   - Do we have comprehensive AI policies in place?
   - Are these policies regularly reviewed and updated?

3. **Regular AI Audits** (Score: 0-3 each)
   - Are regular audits conducted to assess AI performance and compliance?
   - Do these audits address fairness and ethical considerations?

Practical Steps for Boards

4. **Education and Training** (Score: 0-3 each)
   - Are board members provided with AI literacy programs?
   - Do these programs cover AI capabilities, risks, and ethical issues?

5. **Clear Reporting Lines** (Score: 0-3 each)
   - Have we established clear reporting mechanisms for AI initiatives?

- Are these mechanisms effective in ensuring timely updates and accountability?

6. **Stakeholder Engagement** (Score: 0-3 each)

    - Do we engage with stakeholders to align AI strategies with their expectations?

    - Is there a process to incorporate stakeholder feedback into AI governance?

Ensuring Ethical AI Practices

7. **Bias Mitigation** (Score: 0-3 each)

    - Are strategies in place to identify and mitigate biases in AI systems?

    - Is there continuous monitoring to detect and address biases?

8. **Transparency** (Score: 0-3 each)

    - Do we ensure transparency in AI operations and decision-making processes?

    - Are stakeholders informed about AI use and its implications?

9. **Privacy Protection** (Score: 0-3 each)

    - Are stringent privacy standards upheld to protect user data?

    - Is there a framework to handle data breaches effectively?

Monitoring and Evaluation

10. **Continuous Monitoring** (Score: 0-3 each)
    - Is there regular monitoring of AI systems for ethical issues?
    - Are there mechanisms to address identified issues promptly?

11. **Impact Assessments** (Score: 0-3 each)
    - Do we conduct impact assessments to evaluate the ethical implications of AI applications?
    - Are these assessments thorough and actionable?

12. **Stakeholder Feedback** (Score: 0-3 each)
    - Is feedback from stakeholders regularly sought and incorporated?
    - Are improvements made based on this feedback?

Understanding AI's Role in Corporate Strategy

13. **Enhanced Decision-Making** (Score: 0-3 each)
    - Does AI enhance decision-making through sophisticated data analysis?
    - Are AI-driven insights used to refine financial projections and guide capital allocation?

14. **Risk Management** (Score: 0-3 each)

    - Does AI help foresee potential risks and aid in proactive risk mitigation?
    - Are AI tools used to predict outcomes and inform risk management strategies?

15. **Operational Efficiency** (Score: 0-3 each)

    - Does AI streamline information and reporting systems?
    - Are issues identified and addressed in a timely manner?

Establishing Effective AI Governance

16. **Strategic Disruption and Opportunities** (Score: 0-3 each)

    - Are we evaluating how AI could disrupt the industry and what strategic opportunities it presents?
    - Is there a proactive approach to leveraging these opportunities?

17. **AI Utilization** (Score: 0-3 each)

    - Do we understand the role of AI in company processes and third-party products?
    - Is AI usage aligned with business goals and strategies?

18. **Data Governance** (Score: 0-3 each)

    - Are robust data governance practices in place?
    - Do we manage data risks effectively?

19. **AI Governance Systems** (Score: 0-3 each)

    - Is there an effective AI governance framework that integrates insights from all relevant business functions?
    - Are these systems regularly reviewed and updated?

Evaluating AI Projects

20. **Strategic Alignment** (Score: 0-3 each)

    - Do AI initiatives align with the company's strategic goals?
    - Is there a clear strategy for AI deployment?

21. **Resource Assessment** (Score: 0-3 each)

    - Do we have the necessary expertise and resources to pursue AI responsibly?
    - Are resources allocated effectively to AI projects?

22. **Resilience and Compliance** (Score: 0-3 each)

    - Are AI initiatives resilient in terms of cybersecurity and operational integrity?
    - Is there compliance with relevant regulations?

CHAPTER 28 GOVERNANCE AND OVERSIGHT OF AI SYSTEMS

23. **Success Metrics** (Score: 0-3 each)

    - Are clear metrics defined for measuring the success of AI projects?

    - Are proof-of-concept protocols established and followed?

Proactive Oversight Actions

24. **AI Education** (Score: 0-3 each)

    - Are ongoing AI education programs available for board members?

    - Do these programs address current AI trends and best practices?

25. **Developing AI Strategies** (Score: 0-3 each)

    - Are clear AI strategies crafted that align with corporate goals?

    - Is there a focus on responsible AI use and risk management?

26. **Regular Risk Assessments** (Score: 0-3 each)

    - Are regular risk assessments conducted to identify potential AI-related threats?

    - Are these assessments thorough and actionable?

27. **Third-Party Audits** (Score: 0-3 each)

    - Do we engage third parties to audit AI systems and identify vulnerabilities?

    - Are audit findings used to improve AI governance?

CHAPTER 28   GOVERNANCE AND OVERSIGHT OF AI SYSTEMS

28. **Crisis Management Planning** (Score: 0-3 each)

    - Is there a crisis management plan for potential AI-related issues?
    - Are plans regularly reviewed and tested?

29. **External Collaboration** (Score: 0-3 each)

    - Do we consult with AI experts, legal advisors, and consultants to stay updated on emerging AI risks and mitigation strategies?
    - Is external advice integrated into our AI governance framework?

Mitigating Directors' and Officers' Liability

30. **Compliance and Ethics** (Score: 0-3 each)

    - Are AI applications compliant with ethical standards and legal regulations?
    - Is there a process for continuous compliance monitoring?

31. **Data Protection** (Score: 0-3 each)

    - Are robust data protection measures implemented to prevent breaches?
    - Do we ensure compliance with privacy laws?

32. **Documentation of Oversight** (Score: 0-3 each)

    - Are AI oversight activities documented in board minutes and presentations?
    - Is documentation thorough and transparent?

CHAPTER 28 GOVERNANCE AND OVERSIGHT OF AI SYSTEMS

Total Scoring

Each question is scored from 0 to 3, with a maximum possible score of 96 across all 32 questions.

Scoring and Interpretation

- 0–32: High risk – Immediate action required to address significant gaps in AI governance and oversight.
- 33–64: Moderate risk – Areas for improvement identified, and a structured plan is needed to enhance AI governance.
- 65–96: Low risk – Robust AI governance and oversight in place, but continuous monitoring and minor adjustments are recommended.

Threshold for Passing

Organizations should aim for a minimum score of **64** to ensure they have adequately addressed key AI governance considerations and are prepared to manage and scale AI operations effectively.

Summary

In the AI-driven corporate landscape, governance isn't just a checkbox – it's the compass that guides organizations through the ethical minefields and strategic opportunities of artificial intelligence. From establishing robust AI committees to crafting crisis management plans, boards must navigate a complex web of responsibilities that demand both technological savvy and ethical fortitude. As AI systems become more deeply entrenched in business operations, those who master the art of proactive oversight – balancing innovation with risk mitigation, and efficiency with ethical considerations – will find themselves not just compliant, but competitively advantaged. In this era, where AI can be both a transformative ally and a potential liability, will your governance framework be the fortress that safeguards your organization's integrity and propels its AI-powered growth, or the Achilles' heel that leaves you vulnerable to technological, ethical, and legal pitfalls?

CHAPTER 29

Assessing and Advancing AI Maturity in Organizations

Stages of AI Maturity

AI maturity can be assessed across four distinct stages: AI Unaware, AI Aware, AI Ready, and AI Competent. Each stage reflects the organization's understanding, capability, and strategic approach to AI adoption and implementation.

CHAPTER 29   ASSESSING AND ADVANCING AI MATURITY IN ORGANIZATIONS

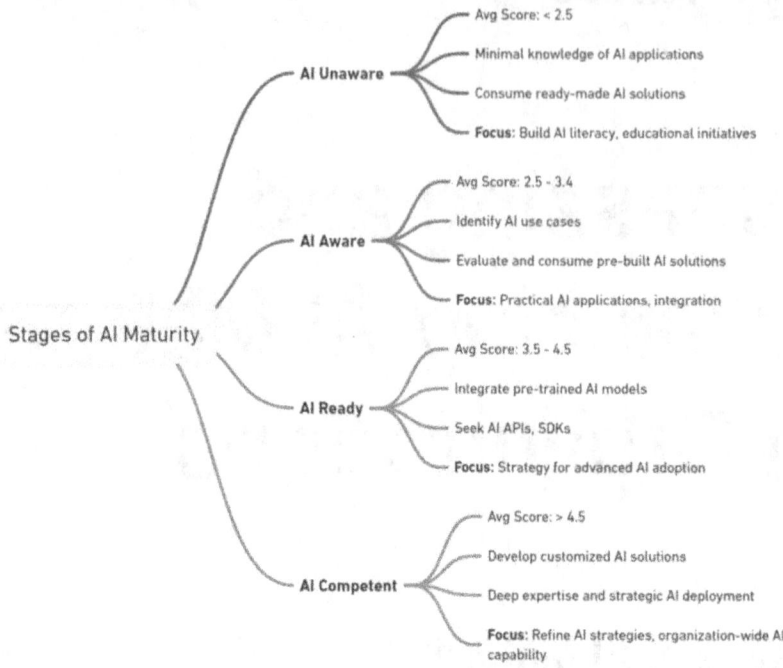

AI Unaware

- **Average Score**: Less than 2.5
- **Interpretation**: Organizations at this stage have minimal knowledge of AI applications and their potential benefits. They might have heard about AI but lack awareness of its practical applications.
- **Characteristics**: These organizations typically wait for vendors to introduce use cases and explain the business value of AI. They predominantly consume ready-made AI solutions.

- **Recommendation**: The primary focus should be on increasing AI literacy across the organization. Engage in educational initiatives and workshops to build foundational knowledge of AI technologies and their business implications.

AI Aware

- **Average Score**: 2.5 to 3.4
- **Interpretation**: Organizations are aware of AI applications and can identify potential use cases within their business.
- **Characteristics**: Actively seeking AI solutions to address business needs, these organizations evaluate and consume pre-built AI solutions. They can identify potential use cases for AI applications.
- **Recommendation**: Broaden the understanding of AI across the organization and start consuming ready-made, end-to-end AI solutions. Initiatives should focus on practical application and integration of AI into existing processes.

AI Ready

- **Average Score**: 3.5 to 4.5
- **Interpretation**: Organizations have the capabilities to integrate pre-trained AI models into their products or business processes.

- **Characteristics**: These organizations evaluate and seek AI APIs, SDKs, and pre-trained models for business use. They are ready to adapt and integrate AI solutions to enhance operational efficiency.

- **Recommendation**: Prepare the organization for advanced AI adoption. Develop a robust strategy to integrate AI solutions and broaden AI understanding to encompass the entire organization.

AI Competent

- **Average Score**: Greater than 4.5

- **Interpretation**: Organizations possess the capabilities to develop customized AI models and solutions tailored to specific business needs.

- **Characteristics**: Having a strategy and roadmap for AI deployment, these organizations can build and implement bespoke AI solutions. They have deep expertise and a comprehensive approach to AI deployment.

- **Recommendation**: Deepen organizational AI capabilities. Focus on refining and optimizing AI strategies, ensuring the entire organization understands and is capable of leveraging AI technologies effectively.

Advancing Through AI Maturity Stages

Transitioning from one stage of AI maturity to the next requires a strategic approach. Here's a detailed roadmap for advancing through each stage:

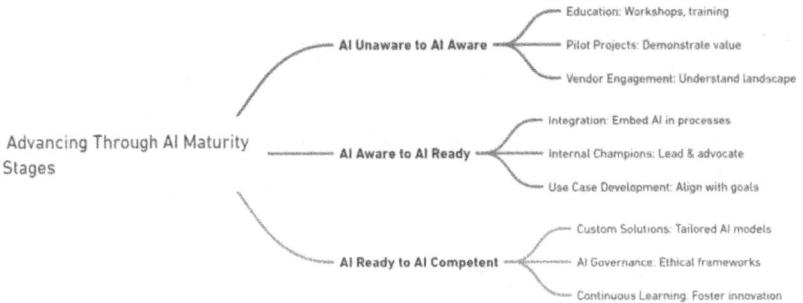

From AI Unaware to AI Aware

- **Education**: Conduct workshops and training sessions to build foundational AI knowledge.

- **Pilot Projects**: Implement small-scale AI projects to demonstrate value and build confidence.

- **Vendor Engagement**: Collaborate with AI vendors to understand the landscape and available solutions.

From AI Aware to AI Ready

- **Integration**: Focus on integrating ready-made AI solutions into existing business processes.

- **Internal Champions**: Identify and train internal AI champions to lead initiatives and advocate for AI adoption.

- **Use Case Development**: Develop a portfolio of AI use cases aligned with business goals.

CHAPTER 29 ASSESSING AND ADVANCING AI MATURITY IN ORGANIZATIONS

From AI Ready to AI Competent

- **Custom Solutions**: Invest in building customized AI models tailored to specific business needs.
- **AI Governance**: Establish robust AI governance frameworks to ensure ethical and responsible AI use.
- **Continuous Learning**: Foster a culture of continuous learning and innovation in AI technologies.

Strategic Considerations for Boards

Effective governance and oversight are critical for successful AI adoption. Boards should play an active role in guiding the organization's AI strategy, ensuring alignment with business objectives and ethical standards.

Key Responsibilities of Boards

- **AI Literacy**: Ensure board members have a fundamental understanding of AI technologies and their implications.
- **Strategic Alignment**: Align AI initiatives with the company's strategic goals and objectives.

- **Risk Management**: Identify and mitigate potential risks associated with AI, including data privacy, security, and ethical concerns.

- **Ethical Oversight**: Establish ethical guidelines for AI use and ensure compliance with regulatory standards.

- **Performance Monitoring**: Regularly review AI projects and their impact on business performance.

Summary

Advancing through the stages of AI maturity is a journey that requires strategic planning, continuous learning, and robust governance. By understanding their current maturity level and implementing targeted strategies, organizations can effectively harness the power of AI to drive innovation, improve efficiency, and achieve their business goals.

Index

A

Accuracy metrics, 154, 166, 167
Action lawsuit, 52
Adaptation and tuning, 262, 270
Adequate resources, 146
Advertising Standards Authority, 31
AI, *see* Artificial Intelligence (AI)
AIaaS, *see* AI-as-a-Service (AIaaS)
AI adoption, 135
    applications, 142, 144
    clear data governance, 142
    data availability and quality, 141
    data ownership and licensing, 139
    data privacy and compliance, 140, 143
    ethical considerations, 141
    ethical guidelines, 143
    high data quality, 144
    integration challenges, 143
    model robustness, 140
    PPM, 138
    problem-solution fit, 141, 144
    robustness of AI models, 143
    robust security measures, 143
    strategic directions, 142
    technical challenges, 140
    technical expertise, 141, 144
    validation and sanitization of inputs, 140
AI-as-a-Service (AIaaS), 211
    description, 204
    Freemium to Premium, 204
    pay-per-use model, 204
    subscription models, 204
AI-assisted content generation, 49
AI budgets and investments
    automation, 127
    checklist
        AI research and innovation, 131
        AI vendors and partners, 131
        long-term investment plan, 131
        revenue streams, 132
        sustainability and ethical considerations, 132
    decision-making, 127, 128
    products and services, 128
    risk management, 128
    scoring and interpretation, 132, 133
    scoring system, 129
    threshold, 133
    total scoring, 132

INDEX

AI budgets and investments checklist
  adoption and integration costs, 131
  AI infrastructure, 130
  AI talent acquisition and development, 130
  budget allocation, 129
  risk management and compliance, 130
  ROI expectations and monitoring, 130
  strategic alignment with business goals, 129

AI copyright and IP consideration checklist
  authorship and inventorship, 54
  continuous monitoring and improvement, 56
  copyright and data training, 55
  data security and breach response, 56
  general compliance, 54
  legal framework and compliance, 55
  passing threshold, 57
  predictive capabilities and IP strategy, 55
  scoring and interpretation, 56

AI-Created Content, 54

AI deployment and governance framework
  creation phase, 11
  discovery phase, 11
  execution phase, 12
  key stages and activities, 10
  operation phase, 13

AI governance, 357
  approaches
    for federal and state governments, 7, 8
    for organizations, 8, 9
  description, 5
  drivers
    ensuring accountability, 6
    legal and regulatory compliance, 6
  integrated framework, 9

AI governance checklist
  board and C-suite executives accountability and compliance, 24, 25
  AI system layer, 22, 23
  continuous improvement and knowledge flow, 25
  environmental layer, 20, 21
  interpretation, 25, 26
  organizational layer, 21, 22
  scoring system, 20
  total scoring, 25

AI Innovations and their IP dilemmas
  AlphaGo by DeepMind, 50
  Artbreeder's AI creations, 50
  chatbots, 50
  IBM Watson, 50
  OpenAI's GPT-3, 50

AI leadership

INDEX

checklist, AI executive compensation insights
  Europe, 335
  United States, 336
compensation
  financial services dominance, 332
  gender and ethnic disparities, 333
  regional variations within the United States, 333
  team size, 332
compensation insights
  equity compensation, 331
  Europe, base salaries/bonuses, 331
  industry-specific compensation, 332
  region, industry, 331
  United States, base salaries/bonuses, 331
ethnic diversity
  Europe, 334
  United States, 334
gender diversity
  Europe, 333
  United States, 333
key considerations
  competitive compensation packages, 334
  continuous professional development, 335
  diversity and inclusion, 335
  strategic reporting lines, 335
organizational structure and roles
  CDAOs, 330
  Chief Data Scientist, 330
  Europe, reporting structure, 330
  experience and background, 330
  machine learning/AI, 330
  reporting structure, 330
  survey, 329
  United States, reporting structure, 330
professional development and retention, 338
reporting structures, 338
scoring and interpretation, 339
specific compensation components
  base salaries and bonuses, 337
  diversity and inclusion, 338
  equity compensation, 337
  industry-specific compensation, 337
threshold, passing, 339
total score, 339
strategic implications (*see* Strategic implications, AI maturity)
AI maturity stages, 371
  AI aware, 373
  from AI aware to AI ready, 375
  AI competent, 374

INDEX

AI maturity stages (*cont.*)
    advancing through AI maturity stages, 375
    AI ready, 373, 374
    from AI ready to AI competent, 376
        strategic considerations, boards, 376
    AI unaware, 372, 373
    from AI unaware to AI aware, 375
    key responsibilities, boards, 376, 377
AI monetization consideration checklist
    common themes, 210
    monetization models, 211, 212
    passing threshold, 213
    scoring interpretation, 213
    scoring system, 209
    total scoring, 212
AI OKR and KPI
    checklist
        accuracy metrics, 166, 167
        adapting metrics to change, 169
        AI project objectives, 166
        alignment with business objectives, 165
        customer satisfaction metrics, 168
        efficiency metrics, 166
        financial impact metrics, 167
        performance metrics, 167
        regular audits and monitoring, 169
        relevant metrics, 166
        scoring and interpretation, 169
        scoring system, 165
        strategic implementation, 168
        threshold, 170
        total scoring, 169
AI operating model checklist
    accountability and governance, 89
    agile and iterative approach, 89
    alignment, business use cases, 88
    clear roles and responsibilities, 92, 93
    collaboration and integration, 89
    data and IT infrastructure, 90
    data science and AI development, 90
    ethical AI practices, 91
    governance and ethical considerations, 90
    interpretation, 93
    passing threshold, 93
    scoring system, 88
    service delivery and performance monitoring, 91
    total scoring, 93
AI operating model framework

INDEX

AI service delivery
  management, 81
  data and AI governance, 81
  data consumers, 81
  data providers, 79, 80
  data science, 80
  IT operations, 80
  organizing AI-related
    activities, 78
AI operating model
  implementation
  continuous monitoring and
    optimization, 83
  defining roles and
    responsibilities, 82
  ethical AI practices, 84
  fostering collaboration across
    teams, 83
  invest in technology and
    infrastructure, 83
  robust data governance
    policies, 83
AI partnerships
  assign relationship managers, 176
  establishing Committee, 175
  key elements (*see* Effective AI
    partnership's elements)
  deep collection, 172
  strategic alliance audit, 175
  strategic alliance playbook, 175
  strategic importance, 171
AI partnerships and alliances
  change management, 121
  checklist

change management, 124
ecosystem and collaborative
  approach, 124
ethical guidelines, 125
governance framework, 125
organization's talent
  development, 124
responsible AI practices, 125
risk management, 125
strategic alignment, 123
technical proficiency,
  123, 124
cultural and organizational
  shifts, 121
decision-making, 119
ecosystem and collaborative
  approach, 121
ethical considerations, 122
organization's talent
  development, 121
risk management and
  governance, 122
scoring and interpretation, 126
scoring system, 122
strategic alignment, 120
technical proficiency, 120
threshold, 126
total scoring, 125
AI-powered products and
  services, 212
developing new offerings, 206
enhancing existing
  products, 205
market differentiation, 206

383

INDEX

AI privacy governance checklist
  consent and data subject rights, 41, 42
  data security and breaches, 42
  design and data transfer, 43
  general compliance, 40, 41
  interpretation, 45
  ongoing evaluation and improvement, 44
  organizational culture and training, 44
  risk assessment and governance, 42
  scoring system, 40
  total scoring, 44
AI regulations considerations
  approaches, 28, 29
  checklist
    future-proofing and innovation, 35
    general AI regulatory compliance, 32
    government regulations and compliance, 33
    legal and ethical considerations, 33
    passing threshold, 36
    regulatory reporting and documentation, 33
    risk-based framework, 33
    scoring and interpretation, 36
    scoring system, 32
    self-regulation and industry standards, 34
    total scoring, 35
  European regulations, 30
  government regulations, 29
  government's role, regulating technology, 31
  self-regulatory framework, 31, 32
  United States, 30, 31
AI risk management
  ethical and social risks, 277, 278
  governance, 288
  key risk categories/mitigation strategies
    bias and fairness, 290, 291
    compliance, 291
    data security, 290
    misuse and harms, 291
    output quality, 289
    privacy, 290
    transparency, 291
  legal and regulatory risks, 276, 277
  managing risks, 288
  map, 288
  measure, 288
  operational risks, 275, 276
  steps, 287
  strategic framework
    high-risk AI systems, 293
    minimal risk AI systems, 292
    transparency risk AI systems, 293

INDEX

unacceptable risk AI systems, 293
strategic recommendations
  continuous monitor/improvement, 295
  foster cross-functional collaboration, 294
  integrate risk management into AI lifecycle, 294
  risk management culture, 294
AI's business value
  enhancing customer experiences, 164
  quantifying performance, 164
  validating financial returns, 164
AI strategy
  boosting market values, 61
  business strategy, 59, 60
  Buy vs. Build Decision (*see* Buy *vs.* build decision)
  cost savings and efficiency gains, 60
  defining, 59
  legal and security assessments, 75
  mindsets (*see* Mindsets, robust AI strategy)
AI strategy consideration checklist
  AI exploration and innovation, 72
  commitment and long-term investment, 74
  ethical AI deployment, 73
  key AI initiatives, 71, 72
  leadership and communication, 73
  legal and compliance groundwork, 70
  passing threshold, 75
  public communication, 74
  roadmapping, 72
  scoring and interpretation, 75
  scoring system, 69
  strategic framework, AI implementation, 70, 71
AI talent
  attracting, 178, 179
  Booz Allen Hamilton (*see* Booz Allen Hamilton)
  development, 179, 180
  landscape, 177
  retention, 181
  strategy implementation (*see* Effective AI talent strategy implementation)
AI talent strategy checklist
  AI talent development, 196
  AI talent retention, 196, 197
  attracting, 195
  implementation and monitoring, 198–200
  passing threshold, 200
  scoring and interpretation, 200
  scoring system, 194
  strategic considerations, 197, 198
  total scoring, 200

INDEX

Algorithmic Accountability Act, 53
AlphaGo's strategic brilliance, 57
Ambitious goals, 135
Anchor hires, 179, 195
Artificial Intelligence (AI), 1, 27, 62
    AI initiatives, 9, 10, 13–15, 26
        metrics (*see* Metrics)
    capabilities, 135
    cost efficiency and growth, 136
    data analysis, 136
    -driven environment, 138
    -driven targeting, 86
    dual benefits, 136
    -generated content, 52, 53
    -generated videos, 49
    individual privacy, 37
    IP and copyright
        landscape, 51, 52
    in IP strategy, 55
    lifecycle, 9
    lifecycle governance
        framework, 13, 14
    market dominance, 2
    maturity, 95, 97
    monetization trends, 208
    privacy and regulations, 3
    responsible AI practices, 97
    service delivery
        management, 81
    transformation, 3, 176
Artificial Intelligence Act (AIA 2023), 6
Attracting AI talent
    anchor hires, 179
    clear value proposition, 178
    customizing recruiting
        processes, 179
    leverage untapped talent
        pools, 179
Augmented and redefined
    leadership roles
    CEO, 187, 188
    Chief Financial Officers,
        193, 194
    Chief Human Resources
        Officer, 194
    Chief Legal and Privacy
        Officers, 192
    Chief Marketing Officers, 193
    Chief Product Officers, 192, 193
    CIOs, CTOs, CDOs, 191
    COO, 188–190

B

Booz Allen Hamilton
    comprehensive training
        programs, 184
    early adoption and centralized
        teams, 183
    partnerships with educational
        institutions, 183
    proactive talent
        mapping, 183
Budget allocation, 129, 148
Business integration, 114
Buy *vs.* build decision
    advanced LLM pipelines, 68

AI sophistication and
    application, 67
basic LLM integration, 67
customized LLM
    implementation, 68
enterprise-wide LLM
    adoption, 69

C

CEO, *see* Chief Executive
    Officer (CEO)
Center of Excellence (CoE), 78, 85,
    86, 105, 108–111, 114
Centralization, 85
Centralized model, 106–107, 110,
    111, 113, 114
Change management, 222
    checklist
        cultural adaptation, 148
        financial management, 148
        framework, 145
        governance and
            leadership, 146
        lawfulness and
            transparency, 145
        measuring success, 149
        monitoring and
            evaluation, 147
        resource allocation, 146
        risk management, 147
        scoring and
            interpretation, 149
        scoring system, 145
    stakeholder
        engagement, 146
    technology integration, 148
    threshold, 150
    total scoring, 149
    training and
        development, 147
communication and
    training, 138
leadership's role, 137
OCM, 137
people-centric approach, 137
plan, 136
PPM, 138
resistance to change, 137
Chatbots, 50, 142, 144, 154, 249,
    293, 309
Checklist determining AI maturity
    advanced retrieval
        augmentation, 100
    advanced retrieval
        augmentation, FFT, 101
    basic retrieval
        augmentation, 100
    foundational level, 99, 100
    intermediate retrieval
        augmentation, 100
    multiagent systems and
        workflow
        orchestration, 101
    orchestrated agentic
        systems, 101
    passing threshold, 103
    responsible AI practices, 102

INDEX

Checklist determining AI maturity (*cont.*)
  scoring and interpretation, 103
  scoring system, 99
  strategic implications, executives and boards, 102, 103
  total scoring, 103
Checklist for AI Risk Consideration, Board/C-Suite
  fairness, 279
  legal and regulatory compliance, 281
  organizational and cultural integration, 282
  privacy, 278
  safety and performance, 280
  security, 279
  third-party risks, 281
  transparency and explainability, 280
Checklist for aligning AI investments
  advanced analytical tools, 220, 221
  departmental enhancements, 220
  enterprise-wide transformation, 221, 222
  fundamental AI applications, 219
  general considerations, 218, 219
  interpretation, 222
  passing threshold, 223
  scoring system, 218
  total scoring, 222

Chief Executive Officer (CEO), 187, 188
Chief Financial Officers, 193, 194
Chief Human Resources Officer, 194
Chief Information, Technology, and Data Officers (CIOs, CTOs, CDOs)
  competencies and actions, 191
  responsibilities, 191
Chief Legal and Privacy Officers, 192
Chief Marketing Officers, 193
Chief Operating Officer (COO)
  additional strategic focus areas, 190
  competencies and actions, 188
  CROs and CISOs responsibilities, 189
  responsibilities, 188
  updated competencies and actions, 189, 190
Chief Product Officers, 192, 193
Chief Technology Officer (CTO), 330, 335, 342, 344, 349
CI/CD, *see* Continuous integration and delivery (CI/CD)
CoE, *see* Center of Excellence (CoE)
Collaborative culture, 124, 344
Collaborative ecosystem, 182, 198
Communication plan, 9, 146, 222
Consent management, 41
Consulting model, 109, 111, 115
Continuous integration and delivery (CI/CD), 174

INDEX

Continuous learning culture, 180, 196
Continuous monitoring, 15, 36, 44, 56, 57, 75, 83, 92, 93, 103, 117, 126, 133, 147, 149, 169, 200, 213, 222, 229, 234, 236, 245, 268, 273, 283, 285, 295, 312, 325, 326, 339, 351, 355, 364
COO, *see* Chief Operating Officer (COO)
Coordination mechanisms, 115
Copyright registration, 51
Cross-functional leaders, 60
Cross-functional project teams, 83
C-suites, 1, 3, 88, 104, 133, 273, 347
CTO, *see* Chief Technology Officer (CTO)
Cultural adaptation, 148
Cultural readiness, 148, 319, 324
Custom AI solutions
    data confidentiality and security, 205
    enterprise customization, 205
    third-party service providers, 205
Customer retention rates, 158, 160, 168
Customer satisfaction metrics, 168
Customer service quality, 168
Cybersecurity posture, AI systems checklist, Board and C-suite
    access control, 231
    collaboration, security experts, 235
    continuous monitoring and testing, 234
    customer trust and transparency, 232
    data handling risks, 233
    data protection, 231
    hardened infrastructure, 230
    identity and input validation, 233
    incident response planning, 235
    responsible AI, 232
    scoring and interpretation, 236
    scoring system, 230
    security, AI models, 234
    threshold, passing, 236
    total score, 236
    training and awareness, 234
    collaboration with security experts, 229
    continuous monitoring and testing, 229
    incident response planning, 229
    LLM security concerns, 227
    organizational AI security posture, 225–227
    strategies and guardrails, 228
    training and awareness, 229

D

Data and AI development, 77
Data-and AI-driven approach, 3

INDEX

Data and business functions collaboration, 78
Data breach response, 42
Data consumers, 81–83
Data handling and privacy, 124
Data privacy framework, 39
Data privacy law, 38
Data protection laws, 6, 140
Data Protection Officer (DPO), 43
Data providers, 15, 78–80, 82, 83, 90
Data science, 1, 15, 60, 80–83, 87, 90, 108, 141, 144, 183, 240, 243, 254, 330
Data security, 9, 39, 42, 56, 130, 205, 206, 231, 251, 290
Data subjects rights, 42
Data transfers, 43
Data utilization, 221
Decentralized model, 107, 108, 110, 114
Decision-making, 2, 85, 128, 163, 170, 217, 285, 291, 343, 346, 356, 364
DeepMind, 50
DPO, *see* Data Protection Officer (DPO)

E

Ecosystem and collaborative approach, 121, 124
Ecosystem integration, 124
Effective AI partnership's elements
 deep collaboration
  joint planning, 172
  risk and investment sharing, 173
  solutions co-creation, 172
  maintaining control and flexibility, 174
  scalability, interoperability, and reusability, 173, 174
Effective AI talent strategy implementation
 assigning dedicated relationship managers, 185
 conducting talent audit, 185
 emphasize ethical AI, 185
 establishing steering committee, 184
 fostering culture of innovation, 185
 monitor and adapt, 186
 strategic talent playbook development, 185
Efficiency metrics, 153, 155, 166
Emergency response times, 136
Enterprise customization, 205, 211
Enterprise initiatives, 114
Enterprise services, 238, 241, 252
Equity compensation, 331, 337
Ethical AI guidelines, 284
Ethical AI practices, 84, 91, 93, 143, 324, 350, 354, 363
Ethical and social risks
 AI, 277
 bias and discrimination, 277
 impact, employment, 278

INDEX

transparency and explainability, 278
Ethical considerations, 12, 15, 22, 24, 26, 62, 81, 90, 91, 97, 111, 122, 125, 130, 132, 141, 206, 277, 354, 370
Ethnic diversity, 334, 336, 338
European Commission, 31, 53
EU's Artificial Intelligence Act (AIA), 6, 7, 28, 30, 33

F

Factory model, 108, 110, 114
Fast Fourier Transform (FFT), 97, 101
Federal Trade Commission, 8
FFT, *see* Fast Fourier Transform (FFT)
Financial impact metrics, 154, 164, 167
Financial services, AI risks management
    AI-powered chatbot, 284
    legal and compliance strategy, 285
    risk assessment and mitigation, 285
    transparency and customer trust, 285
Financial tracking, 148
Flexible work arrangements, 181, 197
Freemium models, 204, 209, 211

Freemium to Premium, 204, 211
Functional alignment, 114
Functional model, 106, 113

G

GDPR Data Privacy laws, 6
Gender diversity, 333
Generative AI, 1, 28, 141, 171, 186
    checklist
        cost assessment and carbon impact, 310
        create value hypothesis, 306
        define solutions, maximize value, 309
        development, testing, and learning, 310, 311
        scaling and adaptation, 311, 312
        scaling through patterns, 307, 308
        score and interpretation, 312
        scoring system, 305
        select foundational Generative AI tools, 308, 309
        threshold, passing, 312
        total score, 312
        use cases prioritization, 306, 307
    costs assessment and carbon impact
        environmental impact, 303
        evaluation, 302

INDEX

Generative AI (*cont.*)
  create value hypothesis
    benchmark potential value, 299
    short-term *vs.* long-term value, 299
    strategic assessment, 298
  define solutions, maximize value
    incremental solutions, 302
    lateral thinking and patterns, 302
    proprietary data integration, 302
  development, testing, and learning
    controlled deployments, 303
    iterative learning, 303
    reevaluation risks and governance, 304
  scaling and adaptation
    adaptive scaling, 304
    broadening applications, 304
    institutional knowledge, 304
  scaling through patterns
    leveraging patterns, 300
    model refinement, 300
    net-new creation and augmentation, value, 300
  select foundational Generative AI tools
    avoid potential tech debt, 301
    customize foundational models, 301
    evaluate technologies, 301
  use cases prioritization
    assessing value and feasibility, 300
    identify high-impact use cases, 299
    industry-specific applications, 299
Generative AI deployment, case study
  continuous learning and adaptation, 321
  logistics management, 320
  precision agriculture, 320
  predictive maintenance, 320
Generative AI Reference Architecture
  adaptation and tuning, 262
  AI systems, 259
  characteristics, trustworthy AI system, 266–268
  checklist
    adaptation and tuning, 270
    enterprise integration, 272
    evaluation and observability, 271
    governance and responsible AI, 272
    MLOps orchestration, 271
    prompt engineering, 269
    RAG, 270

scoring and interpretation, 272
scoring system, 269
security, privacy, and compliance, 271
serving and orchestrating models, 270
threshold, passing, 273
total score, 272
User Experience (UX), 269
enterprise integration, 264
evaluation and observability, 262
governance and responsible AI, 263
MLOps orchestration, 262
prompt engineering, 261
RAG techniques, 261
security, privacy, and compliance, 263
serving and orchestrating models, 261
strategic implications, businesses
  competitive advantage, 265
  enhanced innovation, 264
  operational efficiency, 265
  risk mitigation, 265
  scalability, 265
trustworthy AI system, 265
user experience (U/X), 260
Generative AI use cases checklist
  feasibility and readiness, 323, 324
  financial impact and business model disruption, 321, 322
  operating model and competitive disruption, 322, 323
  responsible AI, 324, 326
  risk management, 324, 326
  score and interpretation, 326
  score system, 321
  threshold, passing, 326
  total score, 326
  key considerations
    business model disruption, 317
    competitive disruption, 318
    drivers of change, 319
    margin (revenue and cost), 316
    model feasibility, 318
    operating model disruption, 317
    responsible AI deployment, 319, 320
Governance and oversight, AI systems
  AI projects, 358
  AI's role, corporate strategy
    board's oversight role, 356
    strategic opportunities, 356
  boards, 353
  business value applications, 360

393

INDEX

Governance and oversight,
  AI systems (*cont.*)
  checklist
    AI projects, 366, 367
    AI's role, corporate strategy, 364, 365
    effective AI governance, 365, 366
    ethical AI practices, 363
    governance and oversight, 362
    mitigating directors' and officers' liability, 368
    monitoring and evaluation, 364
    practical steps, boards, 362, 363
    proactive oversight actions, 367, 368
    scoring and interpretation, 369
    scoring system, 361
    threshold, passing, 369
    total score, 369
  ethical considerations, boards, 354
  ethical guidelines, 355
  governance framework, 358
  governance strategies, 353, 354
  monitoring and evaluation, 355
  practical steps, AI oversight
    mitigating directors' and officers' liability, 360
  proactive oversight actions, 359
  practical steps, boards, 354
  strategic considerations, boards
    key board responsibilities, 361
Governance framework, 13, 112, 115, 125, 146, 168

H

Human-centered design, 136
Human intervention, 153, 166

I, J

IBM Watson's patent insight tools, 50
Independent Software Vendor (ISV), 139
Industry-specific compensation, 332, 337
Insurance, risk management tool, 207
Intellectual Property (IP), 173, 206
  boundaries, 174
  and Copyright laws, 48
  management tools, 55
  operations, 80
Interoperability, 173–176
IP, *see* Intellectual Property (IP)
ISV, *see* Independent Software Vendor (ISV)

INDEX

K

Key performance indicators (KPIs), 19, 60
  AI impact metrics
    customer satisfaction, 157, 158
    operational efficiency, 157
    revenue growth, 158
  defined, 156
  and metrics, 152, 153
    business value, 163, 164
    measurement approach, 165
    refining and guiding future, 164, 165
    and OKR, 165–170
KPIs, *see* Key performance indicators (KPIs)

L

Large Language Model Operations (LLMOps), 97, 101
Large Language Models (LLMs), 97, 226, 227
  applications, 249
  challenges and future directions
    computational resources, 251
    data security, 251
    explainability and transparency, 251
  checklist
    human resources and expertise, 254
    integration systems, expanded AI scope, 255
    investments, 253
    minimum score to pass, 257
    scoring system, 252
    scoring threshold, 257
    technology infrastructure, 252
    total scoring and interpretation, 257
    unique, task-specific enterprise data, 255
  components, 247
  evaluation metrics, 250, 251
  providers, 38
  optimization LLM performance, 249, 250
  parameter file, 247
  security and risks, 249
  training, 248
  transformer architecture, 248
Lawfulness and transparency, 40, 54, 145
Leadership commitment, 146
Legal and ethical considerations
  data privacy and security, 206
  IP rights, 206
  liability and indemnification, 207
Legal and regulatory compliance, 6, 281
Legal and regulatory risks
  compliance with industry standards, 277
  liability and accountability, 277

INDEX

Levels of AI maturity
   advanced retrieval
      augmentation, 96
   advanced retrieval
      augmentation, FFT, 97
   basic retrieval augmentation, 96
   foundational level, 96
   intermediate retrieval
      augmentation, 96
   multiagent systems and
      workflow orchestration, 97
   orchestrated agentic systems, 97
Lifecycle management, 114, 141, 144
LLMOps, *see* Large Language Model Operations (LLMOps)
LLMs, *see* Large Language Models (LLMs)
Local business functions, 78

M

Machine Learning Operations (MLOps), 262
Machine learning techniques, 238, 242, 252
Measurement challenges, 163
Measuring AI success
   adapting metrics to change, 163
   challenges, 160, 161
   data accessibility, 162
   data complexity, 161
   data consistency, 161
   data quality, 161
   external fluctuations, 162
   internal changes, 162
Metrics
   accuracy metrics, 154
   efficiency metrics, 153
   financial impact metrics, 154
   framework for accountability, 152
   and KPIs, 152, 153
   performance metrics, 154
   right metrics, 154–156
Mindsets, robust AI strategy
   AI exploration and innovation, 65
   AI initiatives, 64, 65
   AI strategy roadmapping, 65
   ethical AI deployment, 66
   legal and compliance groundwork, 63
   long-term investment, 62
   strategic framework, AI implementation, 64
   team support, 62
Mitigation AI risks, strategies
   clear legal agreements, AI vendors/partners, 283
   continuous monitoring and auditing, 283
   data governance, 284
   ethical AI guidelines, 284
   quality control, 284
   robust risk management framework, 283

INDEX

transparent and explainable AI, 284
workforce reskilling and transition, 284
MLOps, *see* Machine Learning Operations (MLOps)
Models for structuring AI teams
  advantages and challenges, 105
  align structure with strategic goals, 110
  availability and specialization, 110
  balance technical and business priorities, 111
  centralized model, 106
  checklist
    centralized model, 113
    CoE, 114
    collaboration and innovation, 116
    consulting model, 115
    decentralized model, 114
    factory model, 114
    functional model, 113
    governance and coordination, 115
    monitoring and optimization, 116
    scoring and interpretation, 117
    scoring system, 113
    strategic considerations, 115
    strategic implementation, 116
    threshold, 117
    total scoring, 117
  clear roles and responsibilities, 112
  CoE, 108, 109
  collaboration and innovation, 111
  consulting model, 109
  cross-functional collaboration, 112
  decentralized model, 107, 108
  factory model, 108
  functional model, 106
  governance and coordination, 110
  governance framework, 112
  monitor and optimize performance, 112
  strategic considerations, 109
  training and development, 112

N

NIST, *see* National Institute of Standards and Technology (NIST)
National Institute of Standards and Technology (NIST), 8

O

OCM, *see* Organizational change management (OCM)
Office of Management and Budget (OMB), 30

INDEX

OMB, *see* Office of Management and Budget (OMB)
OpenAI, 38, 208
   AI monetization, 208
   freemium model, 209
   subscription and pay-per-use models, 208
   token-based pricing, 209
Open-source LLM models, 37, 38
Operating model
   challenges, 86, 87
   checklist (*see* AI operating model checklist)
   data and AI integration, 77
   foundation, 78
   non-negotiable element, 77
Operational efficiency, 157
Operational risks
   data quality and integrity, 276
   dependency, external vendors, 276
   impact business continuity and performance, 275
   unpredictable AI behavior, 276
Organizational change management (OCM), 137
Organizational coordination, 114

P

Pay-per-use models, 204, 208, 211
People-centric approach, 137, 150
Performance metrics, 147, 154, 167

Performance monitoring, 91, 116, 174, 210, 280, 361, 377
Portfolio and program management (PPM), 138
PPM, *see* Portfolio and program management (PPM)
Privacy, 45, 290
   by design, 43
   steps
      accountability, 39
      consent, 39
      data minimization, 39
      data security, 39
      transparency, 39
Problem–investment matrix
   advanced analytical tools, 217
   definition, 215
   departmental enhancements, 216, 217
   enterprise-wide transformation, 217, 218
   fundamental AI applications, 216
Productionization, 87
Prompt engineering, 250, 261, 269
Purpose-driven work, 181, 196

Q

Quantifying AI's ROI
   cost savings *vs.* investment costs
      calculating ROI, 160
      direct and indirect costs, 159
      operational cost savings, 160

revenue enhancements, 160
tangible and intangible benefits, 159

R

RAG, *see* Retrieval Augmentation Generation (RAG)
Reporting structures, AI executives
    broad patterns, 342, 343
    CEO, 348
    checklist, scoring system, 347
    common themes, 347, 348
    COO/CAO, 349
    CTO/CIO, 349
    score and interpretation, 350, 351
    strategic directions, executives/boards
        align reporting structures with strategic objectives, 343
        collaborative culture, 344
        empower AI executives, resources and authority, 344
        promote ethical and responsible AI, 345
    strategic recommendations, 349, 350
        cross-functional integration, 346
        empowerment and authority, 346
        ethical oversight, 346
        resource allocation, 346
        strategic alignment, 346
    threshold, passing, 351
    total score, 350
Reskilling programs, 180, 196
Resource allocation, 17, 113, 138, 146, 186, 218, 221, 343, 346, 350
Resource consolidation, 113
Resource specialization, 115
Resource utilization rates, 153, 166
Response times, 136, 154, 157, 164, 167, 168, 217
Retrieval Augmentation Generation (RAG), 261, 270, 301, 308
Return on investment (ROI), 154, 167
    calculation, 149
    quantifying AI, 159, 160
Reusability, 173, 175
Revenue contribution, 167
Revenue growth, 156, 158, 160
Right metrics
    AI project type, 155
    align metrics with business objectives, 155
    balance leading and lagging indicators, 155
    benchmark against industry standards, 156
Risk assessment, 42, 147, 285, 359, 367

INDEX

Risk management, 125, 128, 130, 147, 210, 282, 283, 286–288, 294, 295, 319, 324, 325, 332, 356, 358, 365, 367, 377
Risk management and governance, 122, 125
Risk mitigation, 9, 85, 122, 147, 265, 288, 356, 361, 365
Robust operational AI governance framework
    build phase, 18
    design phase, 17
    run phase, 19
Robust risk management framework, 283
ROI, *see* Return on investment (ROI)

S

Scalability, 9, 65, 72, 85, 102, 130, 165, 173, 175, 176, 208, 264, 265, 300, 301, 307
Scale AI operations, enterprise infrastructure
    checklist
        human resources and expertise, 243
        integration systems, 243
        investments, 242
        scoring and interpretation, 245
        scoring system, 241
        technology infrastructure, 241, 242
        threshold, passing, 245
        total score, 244
        unique, labeled enterprise data, 244
    human resources and expertise, 239, 240
    integration systems, 240
    investments, 239
    technology infrastructure, 238
    unique, labeled enterprise data, 240, 241
Scoring system, 20, 32, 40, 69, 88, 99, 113, 122, 129, 145, 165, 194, 209, 218, 230, 241, 252, 269, 305, 321, 335, 347, 361
Security and privacy, 188, 194, 210, 238, 242, 253, 281, 320, 325
Self-regulatory bodies, 31
Self-regulatory framework, 31, 32, 34
Service quality KPIs, 158
Serving and orchestrating models, 261, 270
Skills assessment, 147
Stakeholder identification, 146
Strategic alignment, 21, 88, 114, 120, 123, 129, 164, 210, 218, 343, 346, 349–351, 358, 366, 376
Strategic approach, AI talent management
    collaborative ecosystem, 182

data-driven decisions, 182
ethical and responsible AI, 182
long-term vision, 183
Strategic implementation, AI deployment and governance
aligning AI initiatives, business strategy, 16
continuous monitoring and improvement, 15
establishing clear governance policies, 14
ethical considerations, 15
fostering cross-functional collaboration, 15
robust data management practices, 15
Strategic implications, AI maturity checklist (*see* Checklist determining AI maturity)
continuous evaluation and optimization, 99
data infrastructure, 98
develop incremental AI capabilities, 98
fostering cross-functional collaboration, 98
prioritizing responsible AI practices, 99
Strategic implications, operating model
alignment with strategic goals, 84
competitive advantage, 85
enhanced decision-making, 85
operational efficiency, 85
risk mitigation, 85
Strategic KPIs, 16
Structured career paths, 180, 185, 196, 199
Subscription-based AIaaS, 204
Subscription models, 204, 208, 211

T

Technical expertise, 123, 141, 144
Technical proficiency, 120, 123, 124
Technology readiness, 148
Third-party risks, 281
Third-party service providers, 205, 212
Traditional brick-and-mortar companies, 2
Transformer architecture, 248
Transparent and explainable AI, 284
Trustworthy AI system, 263, 265, 266, 268, 354

U

US Copyright Office, 51
User Experience (UX), 205, 212, 260, 269, 273
UX, *see* User Experience (UX)

V, W, X, Y, Z

Validation and testing processes, 120, 124

GPSR Compliance

The European Union's (EU) General Product Safety Regulation (GPSR) is a set of rules that requires consumer products to be safe and our obligations to ensure this.

If you have any concerns about our products, you can contact us on

ProductSafety@springernature.com

In case Publisher is established outside the EU, the EU authorized representative is:

Springer Nature Customer Service Center GmbH
Europaplatz 3
69115 Heidelberg, Germany

www.ingramcontent.com/pod-product-compliance
Lightning Source LLC
LaVergne TN
LVHW010333260326
834688LV00036B/692